New Technologies and Concepts for Reducing Drug Toxicities

New Technologies and Concepts for Reducing Drug Toxicities

Edited by
Harry Salem, Ph.D.
Chief, Life Sciences Department
Chemical Research
Development and Engineering Center
Aberdeen Proving Ground, Maryland

Steven I. Baskin, Ph.D.
Pharmacology Division
U.S. Army Medical Research Institute
of Chemical Defense
Department of Physiology
Aberdeen Proving Ground, Maryland

CRC Press
Boca Raton Ann Arbor London Tokyo

Library of Congress Cataloging-in-Publication Data

New technologies and concepts for reducing drug toxicities / Harry
Salem and Steven I. Baskin
p. cm.
Includes bibliographical references and index.
ISBN 0-8493-8896-1
1. Drugs—Toxicology. 2. Drugs—Safety measures. I. Salem,
Harry, 1929- . II. Baskin, Steven I.
RA1238.N48 1993
615′.704—dc20 92-539
CIP

Developed by Telford Press

Direct all inquiries to CRC Press, Inc. 2000 Corporate Blvd., N.W., Boca Raton, Florida, 33431.

International Standard Book Number 0-8493-8896-1

Library of Congress Card Number 92-539

Printed in the United States of America 1 2 3 4 5 6 7 8 9 0

Printed on acid-free paper

Preface

Development of safer drugs not only resides in the discovery of new chemical entities, but also in the application of innovative technologies which enhance the efficacy and reduce the toxicity of therapeutic agents. To help researchers achieve this desired goal, we sought to provide a diversity of concepts and technologies that would allow for the more rational utilization of pharmacologically useful substances.

The authors of the chapters contribute an overview of techniques to improve the effectiveness and safety of the drugs of today and tomorrow.

We thank the authors, Association of Government Toxicologists, publishers, and the family of scientists who have helped to make this book a reality.

Most of the chapters of this book were based on presentations made at a symposium sponsored by the Association of Government Toxicologists. These authors have updated their chapters to encompass related advances achieved since the symposium, and chapters have been added for the sake of completeness and to provide a more extensive overview of new technologies and concepts for enhancing therapeutic efficacy and for reducing drug toxicities.

In addition to the editors of this book, the program committee for the symposium included Drs. Theodore Farber, Sidney Green, Robert Osterberg, John Quest, Esther Rinde, and Sidney Stoltzenberg who contributed to the overwhelming success of the symposium. Special thanks is extended to this committee.

Harry Salem
Steven I. Baskin

The Editors

Dr. Harry Salem is Chief of the Toxicology Division, Chemical Research, Development and Engineering Center, Aberdeen Proving Ground, MD. His research interests and experience are in inhalation and general pharmacology and toxicology. He has served on the editorial boards of several professional journals and is currently a member of the Editorial Board of the Journal of Inhalation Toxicology, and the Editor-in-Chief of the Journal of Applied Toxicology. Dr. Salem has served as Associate Professor of Pharmacology at the University of Pennsylvania, Adjunct Professor of Environmental Health at Temple University and is currently Adjunct Professor at Drexel University.

Dr. Salem was also employed by small and large industry in pharmacology and toxicology. He has served and serves on national and international committees and is Chairman of the Technical Committee for the Inhalation Specialty Section of the Society of Toxicology. His professional affiliations include: American Association for the Advancement of Science, American Chemical Society, American College of Toxicology (Charter Member), American Conference of Governmental Industrial Hygienists, Inc., American Society of Clinical Pharmacology and Therapeutics, American Society for Pharmacology and Experimental Therapeutics, Association of Governmental Toxicologists where he served as President, Chemical Corps Association Inc., European Society of Toxicology, International Regulatory Pharmacology and Toxicology, International Society on Toxinology, National Capitol Area Chapter Society of Toxicology, Society of Toxicology, and Sigma Xi.

He serves on the Steering Committee for the Society of Comparative Ophthalmology, and is a Fellow of the New York Academy of Sciences, the American College of Clinical Pharmacology, and the Academy of Toxicological Sciences where he served on the Professional Standards Evaluation Board, and currently serves on the Board of Directors.

He has contributed many scientific papers and is the co-editor of three volumes of the *International Encyclopedia of Pharmacology and Therapeutics on Antitussive Agents,* and is the editor of the text, *Inhalation Toxicology.* Dr. Salem received the B.A. degree (1950) from the University of Western Ontario in Canada, B.Sc. degree (1953) in pharmacy from the University of Michigan, and M.A. (1955) and Ph.D. (1958) degrees in pharmacology from the University of Toronto in Canada.

Steven I. Baskin was born and raised in Los Angeles, CA. He attended and received a Doctor of Pharmacy (Pharm.D.) from University of Southern California in 1966, a Ph.D. in pharmacology from Ohio State University in 1971, and completed a post doctoral training from Michigan State University in 1973 with Dr. Ted Brody. From 1973 to 1982, he was a faculty member in the Pharmacology Department at Medical College of Pennsylvania, and for the last 10 years he has been with the United States Medical Research Institute of Chemical Defense. Currently, he is team leader of Biochemical Toxicology. He has co-authored approximately 100 papers and co-edited four books.

Dr. Baskin is a member of ACS, SOT, ASPET, AGT, ABT, Sigma Xi, ACC, Cardiac Muscle Society, and other related scientific organizations. He has held positions in ASPET and AGT. He is the treasurer of AGT and an editor of the Journal of Applied Toxicology. His research interests include cardiovascular toxicology and mechanisms of cyanide and sulfur-containing substances.

Contributors

Abraham Abuchowski
Enzon, Inc.
South Plainfield, NJ

Steven I. Baskin
Pharmacology Division
U.S. Army Medical Research Institute of Chemical Defense
Aberdeen Proving Ground, MD

Jill M. Bensie
Drug Dosing Services, Inc.
Southfield, MI

Jerry N. Blancato
Exposure Monitoring Branch
Exposure Assessment Research Group
Environmental Monitoring Systems Laboratory
U.S. Environmental Protection Agency
Las Vegas, NV

Henry Brem
Department of Neurosurgery
Johns Hopkins University School of Medicine
Baltimore, MD

Brian P. Brockway
Data Sciences, International
St. Paul, MN

Friedericka Fuertges
Enzon, Inc.
South Plainfield, NJ

Shayne C. Gad
Medical Affairs Technical Services
Becton Dickinson
Research Triangle Park, NC

S. K. George
Department of Medical Biochemistry and Genetics
Texas A&M College of Medicine
College Station, TX

Jim Harlan
Cogitate
Southfield, MI

Craig R. Hassler
Battelle Memorial Institute
Columbus, OH

Garret M. Ihler
Department of Medical Biochemistry
Texas A&M College of Medicine
College Station, TX

Shirley A. Liebman
U.S. Army Chemical Research, Development and Engineering Center
Aberdeen Proving Ground, MD

Donald M. Maxwell
U.S. Army Medical Research Institute of Chemical Defense
Aberdeen Proving Ground, MD

Alessandro Olivi
Department of Neurosurgery
Johns Hopkins University School of Medicine
Baltimore, MD

Mary Prevo
Alza Corporation
Palo Alto, CA

Harry Salem
Life Sciences Department
Chemical Research Development and Engineering Center
Aberdeen Proving Ground, MD

David R. Schneider
Department of Pharmacology
School of Medicine
Wayne State University
Detroit, MI

A. Peter Snyder
U.S. Army Chemical Research, Development and Engineering Center
Aberdeen Proving Ground, MD

Warren C. Stern
Cato Research
Durham, NC

Rafael J. Tamargo
Department of Neurosurgery
Johns Hopkins University School of Medicine
Baltimore, MD

Y.-H. Xu
Department of Medical Biochemistry and Genetics
Texas A&M College of Medicine
College Station, TX

Contents

Dedication

The editors dedicate this book to Flo, Jerry, Amy and Joel as well as to Sue, Lloyd, Miriam, Zvi, and Kendall. It is hoped that this will encourage them to contribute to advancing technologies and concepts for the betterment of all peoples.

Chapter

1

Toxicokinetic Modeling of Organophosphorus Compounds to Reduce Toxicity*

Donald M. Maxwell
U.S. Army Medical Research Institute of Chemical Defense
Aberdeen Proving Ground, Maryland

INTRODUCTION

Phosphorus plays a central role in living organisms. Photosynthesis, energy metabolism, cellular membranes, and replication all require naturally occurring phosphorus compounds. In a similar fashion, synthetic organophosphorus compounds pervade everyday living. Organophosphorus compounds are used in industry (lubricants, antioxidants, flame retardants, plasticizers, corrosion inhibitors, antistatic agents, emulsifiers), agriculture (insecticides, fungicides, herbicides, rodenticides, growth regulators), medicine (radiodiagnostic agents, cholinergic drugs, antitumor drugs), and warfare (nerve agents).

The widespread applications of synthetic organophosphorus compounds lead inevitably to human exposure. The populations at greatest risk for exposure to organophosphorus compounds are chemical workers, industrial workers, agricultural workers, and military personnel. The potential for human exposure to organophosphorus compounds creates the need for (1) human risk

*The opinions or assertations contained herein are the private views of the author and are not to be construed as official or as reflecting the views of the Army or the Department of Defense.

0-8493-8896-1/93/$0.00 + $.50

assessment to establish safe exposure levels and (2) development of biomedical strategies to reduce the toxicity of organophosphorus compounds. Both problems can be addressed by toxicokinetic modeling of appropriate databases.

The major difficulty in providing human risk assessment for organophosphorus compounds is an inadequate database for human exposure. Since human exposures are primarily accidental, exposure dosage as well as blood and tissue concentrations of the organophosphorus compound may be unknown. This inadequate human database necessitates the extrapolation of animal data to human situations. The advantage of data generated in animals is that the required data (i.e., dosage, blood concentrations, toxicological response) can be carefully measured in animals under controlled conditions. In addition, the diverse biochemical reactions of organophosphorus compounds, which occur simultaneously in the body, can be characterized independently by separate *in vitro* experiments.

BIOCHEMISTRY OF ORGANOPHOSPHORUS COMPOUNDS

The general structure of organophosphorus compounds is shown in Table 1. A_1 and A_2 are usually rather unreactive groups, such as alkyl or alkoxy groups. X is a highly reactive group, such as a halide or thiol. The covalent bond between P and X is the most labile of the four bonds attached to phosphorus. Therefore, the P–X bond is preferentially cleaved when organophosphorus compounds react with enzymes.

The biochemical reactions of organophosphorus compounds are shown in Table 2. The biochemical reaction that produces the pharmacological response to an organophosphorus compound (I) is the irreversible inhibition of acetylcholinesterase (AChE). The inhibition of AChE results in the accumulation of acetylcholine, a neurotransmitter normally hydrolyzed by AChE. The accumulation of acetylcholine produces a variety of pharmacological effects, including tremors, muscular weakness, decreased heart rate, and respiratory depression, which, if severe enough, can result in death (Natoff, 1971; Brimblecombe, 1977). Organophosphorus compounds also react with detoxification enzymes, such as organophosphate hydrolase (Hyd) and carboxylesterase (CaE), that can reduce the concentration of organophosphorus compound available to inhibit AChE. Organophosphate Hyd catalyzes the hydrolysis of organophosphorus compounds (Gay and Hoskins, 1979), whereas CaE reacts irreversibly with organophosphorus compounds in a manner analogous to their reaction with AChE. The reversible binding of organophosphorus compounds to macromolecules (i.e., proteins, lipids) is important because this reaction partitions these compounds between tissues and plasma, which can influence their pharmacokinetic distribution. All of the biochemical

TABLE 1
Examples of Common Organophosphorus Compounds

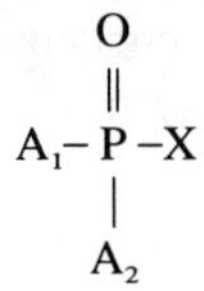

Name	Use	A_1	A_2	X
Dichlorvos	Insecticide	MeO	MeO	$OCH{=}CCl_2$
Paraoxon	Research	EtO	EtO	$OPh{-}4{-}NO_2$
Ecothiophate	Drug	EtO	EtO	$SC_2H_4N^+Me_3$
Soman	Warfare	Me	$Bu^tCH(Me)O$	F
Tritolylphosphate	Lubricant	MePhO	MePhO	OPhMe

Note: Abbreviations: methyl (Me), ethyl (Et), phenyl (Ph), and tertiary butyl (Bu^t).

TABLE 2
Biochemical Reactions of Organophosphorus Compounds

1. Pharmacological	$AChE + I \xrightarrow[k_{AChE}]{} AChE{-}I$
2. Detoxification	$CaE + I \xrightarrow[k_{CaE}]{} CaE{-}I$
3. Detoxification	$I \xrightarrow[k_{Hyd}]{} I'$
4. Reversible binding	$Tissue + I \underset{R^*}{\rightleftharpoons} [Tissue \cdot I]$

* R = [Tissue · I]/[I]

reactions of organophosphorus compounds can be characterized by either reaction rate constants (k_{AChE}, k_{CaE}, k_{Hyd}) or partition coefficients (R) that can be measured in separate *in vitro* experiments.

TOXICOKINETIC MODELING

Based on the mass-balance principles of physiological pharmacokinetics (Gibaldi and Perrier, 1982; Gerlowski and Jain, 1983), the composite effect in a tissue of all of the biochemical reactions of an organophosphorus compound can be described by the following equations:

$$dI_i/dt = Q_iI_p/V_i - Q_iI_i/R_iV_i - I_i\,(k_{AChE}AChE + k_{CaE}CaE + k_{Hyd})$$

$$dAChE_i/dt = -k_{AChE}AChE_iI_i$$

$$dCaE_i/dt = -k_{CaE}CaE_iI_i$$

where:

- I = Organophosphorus compound concentration
- AChE = Acetylcholinesterase concentration
- CaE = Carboxylesterase concentration
- Q = Blood flow
- V = Volume
- R = Partition coefficient
- k = Reaction rate constants
- t = Time
- p = Plasma

dI_i/dt is the change in concentration of an organophosphorus compound in tissue (i) with time. $dAChE_i/dt$ and $dCaE_i/dt$ are the corresponding changes in concentration of AChE and CaE. The enzyme (Hyd) that hydrolyzes organophosphorus compounds is not consumed in the process, and thus $dHyd_i/dt = 0$. The equation describing dI_i/dt is a mathematical expression that the tissue concentration of an organophosphorus compound equals the amount which flows into the tissue (Q_iI_p) minus the amount which flows out (Q_iI_i/R_i) minus the amount which is removed by biochemical reactions (i.e., AChE, CaE, Hyd). If a similar set of equations is written for each major tissue and organ, the resulting system of equations describes the distribution and metabolism of an organophosphorus compound in the body. This system of equations can be solved by specialized mathematical algorithms (Hindmarsh, 1974) to calculate the concentration of I, AChE, or CaE at any time after exposure to an organophosphorus compound.

PHYSIOLOGICAL VS. TRADITIONAL PHARMACOKINETICS

The differences between toxicokinetic models that are based on physiological pharmacokinetics and models that are based on traditional pharmacokinetics are shown in Table 3. Traditional pharmacokinetics describes the time course of the plasma concentrations of a chemical compound by a series of exponential terms, one for each tissue compartment. This mathematical expression is solved by fitting concentrations and time data to yield the coefficients for each exponential term. The mathematical expression for a traditional pharmacokinetic model is restricted to describing one set of data and is not very useful for extrapolating to novel situations. In contrast, a physiological pharmacokinetic model is a series of equations describing the transport, reversible binding, and biochemical reactions of a chemical compound. Blood flows, tissue volumes, partition coefficients, and reaction rate constants that are incorporated into the model are measured independently. The system of equations is solved to predict tissue concentrations resulting

TABLE 3
Comparison of Traditional and Physiological Pharmacokinetics

Traditional	Physiological
$dC/dt = d(Ae^{-at} - Be^{-bt} \ldots)/dt$	$dC_p/dt = Q_iC_i/V_i - Q_pC_p/V_p - k_pC_p$ $dC_i/dt = Q_pC_p/V_p - Q_iC_i/R_iV_i - k_iC_i$ . . .
Input: C,t	Input: Q,V,R,k,t
Output: A,B, . . . a,b, . . .	Output: C_p,C_i . . .

from an input dosage of a chemical compound. The ability to calculate the effect of changes in any input parameter (i.e., tissue volume, blood flow, reaction rate, partition coefficient) makes physiological pharmacokinetic models into powerful tools to extrapolate from existing databases into new situations.

MODEL VALIDATION

Toxicokinetic models are tested by comparing model predictions to experimental data. The change in concentration of either AChE, CaE, I, or all three components can be used to validate the model. In Figures 1 and 2, an example of the comparison between model predictions and experimental data can be examined by comparing the levels of AChE, CaE, and I measured in mice after exposure to soman, a highly toxic organophosphorus compound. Since soman alters CaE in only one tissue, plasma, the change in CaE is not a particularly meaningful validation test for the toxicokinetic model (Figure 1). However, AChE and soman concentrations are altered in all tissues and therefore these parameters constitute a good test of the correspondence between model predictions and experimental measurements (Figure 2). The theoretical predictions in Figures 1 and 2 were considered to agree with experimental values, since they were within one standard deviation of the mean of the experimental measurements.

SENSITIVITY ANALYSIS

After a model is validated, it can be used for human risk assessment or for developing methods to reduce compound toxicity. For human risk assessment, blood flows, tissue volumes, and reaction rates appropriate for humans would be substituted into the model to estimate the concentrations of enzymes and organophosphorus compound after human exposure. For development of strategies to reduce toxicity, a technique called sensitivity analysis is used.

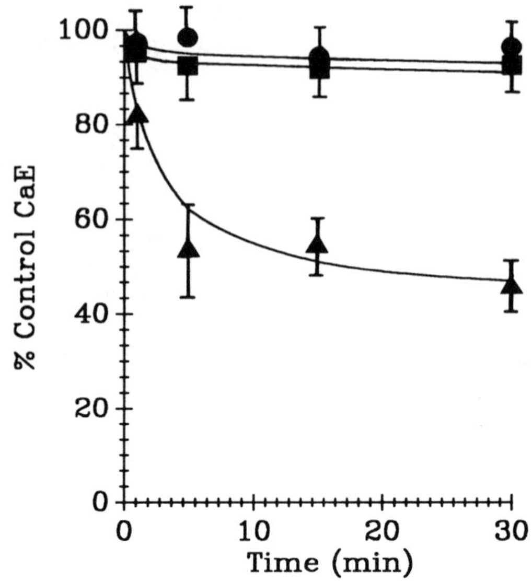

Figure 1. Time course of CaE inhibition following i.v. administration of 25 μg soman to mice. Points represent means ± SD (n = 5) for experimental measurements in diaphragm (●), brain (■), and plasma (▲). Curves represent toxicokinetic model predictions. Data derived from covalently bound soman measurements of Reynolds et al. (1985).

Sensitivity analysis is the estimation of the dependency of the output of a model on changes in the parameters that are in the model. For small systems of equations, formal mathematical algorithms exist, but for large systems incremental sensitivity analysis is used (Carson et al., 1983). Incremental sensitivity analysis is a process by which each parameter in the model is varied one at a time by a chosen increment and computer simulations are used to generate the model output corresponding to these incremented values. Table 4 contains the results of incremental analysis of the sensitivity of brain AChE to 10% increases in a variety of model parameters. Inhibition of brain AChE was chosen as the model output because of its important role in the toxicological response to organophosphorus compounds. The magnitude of the response of brain AChE inhibition to equal variation in each model parameter defined the relative importance of each parameter in producing the model output. The parameters that produced the greatest changes in brain AChE inhibition were tissue blood flow (−3.2%), plasma CaE (+4.7%), and k_{CaE} (+3.9%). k_{Hyd} and the other tissue parameters had no strong effect on the predicted level of brain AChE inhibition. From this incremental sensitivity analysis several theoretical predictions can be made:

1. An increase in tissue blood flow will increase the inhibition of AChE in that tissue.

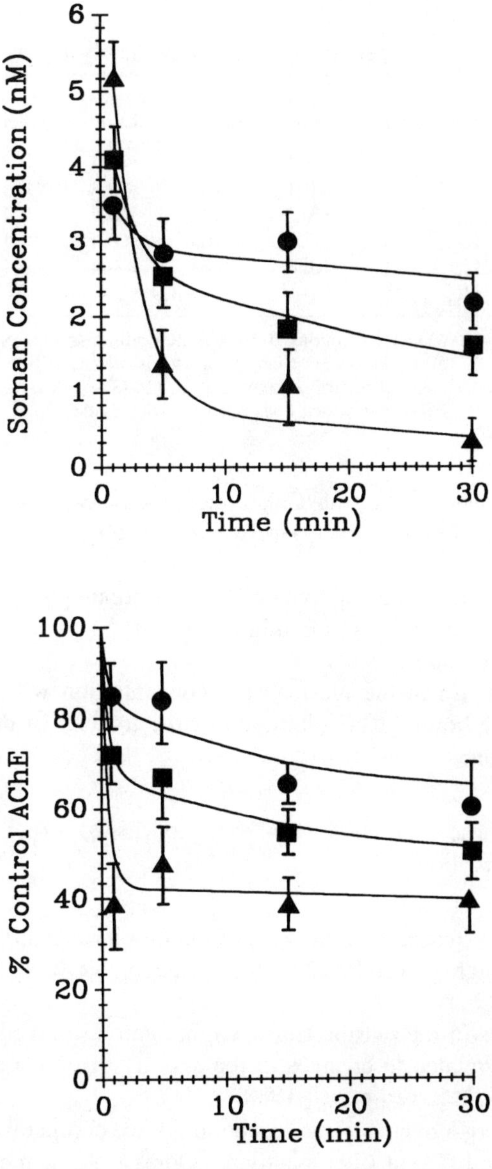

Figure 2. Time course of AChE inhibition and organophosphorus compound concentration following i.v. administration of 25 μg soman to mice. Points represent means ± SD (n = 5) for experimental measurements in diaphragm (●), brain (■), and plasma (▲). Curves represent toxicokinetic model predictions. Data taken from Reynolds et al. (1985).

TABLE 4
Incremental Sensitivity Analysis

Parameter	% Change in brain AChE activity[a] Brain	Plasma	Lung	Liver
[AChE]	+0.8	<0.1	<0.1	<0.1
[CaE]	+1.4	+4.7	<0.1	<0.1
Q	−3.2	<0.1	<0.1	<0.1
R	−0.4	<0.1	<0.1	<0.1
k_{Hyd}	<0.1	<0.1	<0.1	<0.1
k^{b}_{CaE}		+3.9		

[a] Model predictions at 30 min after administration of 50 μg/kg soman when parameter is increased 10%. Data taken from Maxwell et al. (1988).
[b] Parameter which does not vary with tissue.

2. An increase in plasma CaE concentration will reduce brain AChE inhibition and consequently the toxicity of organophosphorus compounds.
3. An increase in the reactivity of CaE for organophosphorus compounds will reduce brain AChE inhibition and the toxicity of organophosphorus compounds.
4. An increase in the plasma Hyd concentration will not significantly alter the brain AChE inhibition or the toxicity of organophosphorus compounds.

Each of these predictions has been confirmed experimentally by the following observations:

1. Tissue differences in the rates of inhibition of AChE have been found to be highly correlated to the tissue blood flow (Maxwell et al., 1987b).
2. Changes in the plasma CaE of aging animals have been found to be closely related to changes in the toxicity of organophosphorus compounds (Maxwell et al., 1988).
3. When organophosphorus compounds have comparable AChE reactivity, but different CaE reactivity (Ooms, 1961), the compound with higher CaE reactivity is less toxic (Maxwell, 1989).
4. Plasma Hyd activity, which varies among species (DeJong et al., 1988), is not correlated with the organophosphorus compound toxicity in these species (Maxwell et al., 1987a).

PRACTICAL APPLICATIONS

Since the *in vivo* effects of hypothetical changes in individual model parameters can be predicted by sensitivity analysis, the most probable sites of beneficial intervention can be suggested. Sensitivity analysis of the toxicokinetic models for organophosphorus compounds suggests that the best biomedical approach to reduce the toxicity of organophosphorus compounds is through the manipulation of the reactivity (k_{CaE}) or concentration of plasma CaE. This general approach has been tested recently in several studies of the ability of plasma scavengers (i.e., CaE, AChE, other esterases) to prevent the toxicity of organophosphorus compounds (Doctor, 1989). Administration of exogenous scavengers has been quite successful in preventing the toxicity of organophosphorus compounds without the side effects that occur when conventional pharmacological treatments (i.e., anticholinergics and AChE reactivators) are used. This success demonstrates that new pharmacological treatments can quickly result from the analysis of appropriate toxicokinetic models.

CONCLUSIONS

The toxicokinetic model for soman and similar models for other organophosphorus compounds (Gearhart et al., 1989) are useful for analyzing several problems associated with the toxicity of these compounds. First, they incorporate and summarize data from a variety of *in vitro* and *in vivo* experiments into a single mathematical model. Second, they provide predictions for human risk assessment by providing a rational framework to extrapolate from animal data to the human situation. Finally, they suggest new concepts or treatments to reduce the toxicity of organophosphorus compounds. The continued refinement of these toxicokinetic models should increase their usefulness for these applications and extend their use to new compounds, exposure scenarios, and applications in the future.

BIBLIOGRAPHY

Brimblecombe, R.W.: Drugs acting on central cholinergic mechanism and affecting respiration. *Pharmacol. Ther. B*. 3:65 (1977).

Carson, E.R., C. Cobell, and L. Finkelstein: *The Mathematical Modeling of Metabolic and Endocrine Systems,* John Wiley & Sons, New York (1983), pp. 227–230.

DeJong, L.P.A., C. VanDijk, and H.P. Benschop: Hydrolysis of the four stereoisomers of soman catalyzed by liver homogenate and plasma from rat, guinea pig and marmoset and by human plasma. *Biochem. Pharmacol.* 37:2939 (1988).

Doctor, B.P.: Enzymes as a scavenger of nerve agents: a feasible concept, in 3rd International Symposium on Protection Against Chemical Warfare Agents, Swedish Defence Research Institute, Umea, Sweden (1989), pp. 225–226.

Gay, D.D. and F.C.G. Hoskins: Stereospecificity and active site requirements in a diisopropylphosphorofluoridate-hydrolyzing enzyme. *Biochem. Pharmacol.* 28:1259 (1979).

Gearhart, J.M., G.W. Jepson, H.J. Clewell, M.E. Anderson, and R.B. Conolly: Physiologically-based pharmacodynamic model for inhibition of acetylcholinesterase (AChE) by diisopropylfluorophosphate (DFP). *Toxicologist* 9:19 (1989).

Gerlowski, L.E. and R.K. Jain: Physiologically based pharmacokinetic modeling: principles and applications. *J. Pharm. Sci.* 72:1103 (1983).

Gibaldi, M. and D. Perrier: *Pharmacokinetics,* 2nd ed., Marcel Dekker, New York (1982), pp. 355–384.

Hindmarsh, A.C.: GEAR: Ordinary Differential Equation System Solver, Lawrence Livermore Laboratory Report UCID-3001, Rev. 3. Lawrence Livermore National Laboratory, Livermore, California (1974).

Maxwell, D.M.: Nerve agent specificity of scavenger protection by carboxylesterase, in 3rd International Symposium on Protection Against Chemical Warfare Agents, Swedish Defence Research Institute, Umea, Sweden (1989), Supplement, pp. 52–57.

Maxwell, D.M., K.M. Brecht, and B.L. O'Neill: The effect of carboxylesterase inhibition on interspecies differences in soman toxicity. *Toxicol. Lett.* 39:35 (1987a).

Maxwell, D.M., D.E. Lenz, W.A. Groff, A. Kaminskis, and H. Froehlich: The effects of blood flow and detoxification on *in vivo* cholinesterase inhibition by soman in rats. *Toxicol. Appl. Pharmacol.* 88:66 (1987b).

Maxwell, D.M., C.P. Vlahacos, and D.E. Lenz: A pharmacodynamic model for soman in the rat. *Toxicol. Lett.* 43:175 (1988).

Natoff, I.L.: Organophosphorus pesticides. *Pharmacol. Prog. Med. Chem.* 8:1–37 (1971).

Ooms, A.J.J.: The Reactivity of Organic Phosphor Combinations in Regards to a Number of Esterases. Ph.D. thesis, University of Leiden, The Netherlands (1961).

Ooms, A.J.J. and J.C.A.E. Breebaart-Hansen: The reaction of organophosphorus compounds with hydrolytic enzymes: the inhibition of horse liver aliesterase. *Biochem. Pharmacol.* 14:1727 (1965).

Reynolds, M.L., P.I. Little, B.F., Bagley, and B.R. Martin: Relationship between the biodisposition of [^{3}H]soman and its pharmacological effects in mice. *Toxicol. Appl. Pharmacol.* 80:409 (1985).

Chapter

2

An Overview: Analytical Toxicology and Integrated Intelligent Instruments (I^3)

Presented at
"New Technologies and Concepts for Reducing Drug Toxicities"
sponsored by the Association of Government Toxicologists
October 26 to 27, 1988

Shirley A. Liebman and A. Peter Snyder
U.S. Army Chemical Research, Development
and Engineering Center
SMCCR-RSL
Aberdeen Proving Ground, Maryland

INTRODUCTION

Developments in the analytical instrumentation industry best demonstrate the effective technology transfer between manufacturers and users throughout industry, government, and academia. One such effort was the Macromolecules in Microencapsulation Workshop hosted by the U.S. Army Chemical Research, Development and Engineering Center on the microencapsulation field which featured process technology, physicochemical research, characterization instrumentation, and methods. This overview will focus upon the latter topic, emphasizing trace organic analysis in toxicology, forensics, environmental, and industrial research and development (R&D). The approach of integrated, microprocessor-based instrumentation is demonstrated in trace organic analysis for target chemicals and biologicals in air, water/biological

0-8493-8896-1/93/$0.00 + $.50

GENERAL PERSPECTIVE

- **MICROPROCESSOR-BASED INSTRUMENTATION AND METHOD DEVELOPMENT**
- **SAMPLE PROCESSING WITH TRACE DETECTION/IDENTIFICATION**
- **CHEMOMETRICS AND APPLIED AI**

SPECIFIC EXPERIMENTATION

ANALYTICAL PYROLYSIS COUPLED TO:

- **CONCENTRATOR GAS CHROMATOGRAPHY**
- **GAS CHROMATOGRAPHY**
- **TANDEM MASS SPECTROMETRY**

SUPERCRITICAL FLUID TECHNOLOGY

EXTRACTION, DESORPTION, REACTION, CHROMATOGRAPHY

CURRENT FOCUS

INTEGRATED INTELLIGENT INSTRUMENTS (I^3) - EMBEDDED INTELLIGENCE AND DECISION-SUPPORT/INTERPRETATION

Figure 1. Instrumentation criteria for polymer characterization and degradation mechanism elucidation.

fluids, and solids. Sampling protocols are included with the concentration/enrichment, separation, and detection/identification stages for complex problem-solving situations. Microencapsulated materials represent complex systems within the general material sciences field, and these polymeric systems are found in a wide range of applications throughout the biomedical, agricultural, pharmaceutical, and food industries. Figure 1 outlines the I^3 approach to the analytical toxicology field with emphasis on selected instrumentation using analytical pyrolysis and supercritical fluid technologies.

ANALYTICAL PYROLYSIS TECHNOLOGY (1969 TO THE PRESENT)

The commercial availability of pyrolyzer systems in 1969 spurred development of analytical pyrolysis methods throughout industrial R&D.[1] Needs for environmental trace analyses were triggered by the vinylchloride monomer problem in 1970. Method development involved pyrolysis units interfaced to a full range of gas chromatographic and spectral systems. In the mid 1970s, automated sample concentrators or enrichment modules were commercially available[2] (Figure 2). The need for advanced sample processing and on-line analysis led to the development of fully automated modular systems with

TRACE ORGANIC ANALYSIS/ENVIRONMENTAL PROBLEMS

ADDRESSED BY ANALYTICAL PYROLYSIS

INSTRUMENTATION TECHNOLOGIES

- **GC** • **MULTIDIMENSIONAL GC** • **MICROPROCESSOR GC**
- **GC-IR** • **FTIR** • **GC-FTIR**
- **MS** • **GC-MS** • **CONCENTRATOR-GC**
- **TGA-FTIR** • **TGA-MS** • **GC-FTIR** • **GC-MS**

Figure 2. Outline of instrumentation for sample processing and on-line analysis (1960s to 1970s). GC, gas chromatography; IR, infrared; FTIR, Fourier transform IR; MS, mass spectrometry; TGA, thermogravimetric analysis.

three distinct functions: sample processing/concentration, separation, and detection/identification, and formed the basis of the I^3 concept (Figure 3).

An illustration of a method to determine polymer microstructure and composition with pyrolysis (Py)-GC is shown in Figure 4. A few tenths of a milligram of a high-molecular weight copolymer of polyethylene and polyvinylacetate is first examined by a dynamic headspace (DHS) method to devolatilize the sample (Figure 4a,b). This analysis determines residual solvents, monomers, solid processing aids, etc., by heating the sample to a relatively low temperature (120°C) for 5 min, trapping the volatiles and then analyzing them by programmed GC (Figure 4a,b). The same sample is then pulse pyrolyzed to 700°C for 20 s in helium to produce a fragmentation pattern that is informative as to the polymer microstructure (Figure 4c,d). Hence, DHS and pyrolysis treatments on a single sample give comprehensive characterization data. Furthermore, if the pyrolysis is conducted under an air atmosphere (Figure 4b,d) rather than helium (Figure 4a,c), oxidative pyrolysis studies may be conducted for information relating to oxidative stability, processing, and/or long-term aging characteristics.

FORENSIC AND ENVIRONMENTAL SCIENCE

The role of analytical pyrolysis has been a dominant one in forensic science for the trace analysis of microorganisms, toxins, arson, and crime scene evidence such as fibers, paints, and biologicals.[3] Likewise, the pyrolyzer/concentrator sample processing units are used in environmental

SAMPLE PROCESSING	**THERMAL TREATMENTS** **PULSE, PROGRAMMED PYROLYSIS** **CONTROLLED ATMOSPHERES**
CONCENTRATION	**CONCENTRATOR TECHNOLOGY** **SORBENT AND CRYOGENIC ENRICHMENT** **DYNAMIC HEADSPACE**
SEPARATION	**PACKED, FUSED SILICA CAPILLARY** **MULTIDIMENSIONAL GC COLUMNS**
DETECTION	**GC DETECTORS** **TCD, FID, PID, NPD, FPD, ECD** **SPECTRAL DETECTORS** **FTIR, MS, MS/MS**
DATA ANALYSIS	**CHEMOMETRICS** **ARTIFICIAL INTELLIGENCE (AI)**

Figure 3. Outline of fully automated instrumentation for advanced sample processing and analysis (1970s to 1980s). TCD, thermal conductivity; FID, flame ionization; PID, photoionization; NPD, nitrogen-phosphorus; FPD, flame photometric; ECD, electron capture.

applications to analyze priority pollutants in air, water, and solids with the aid of automated sample concentrators.[2] In such configurations, sorbent bed trapping cartridges were desorbed initially with carbon disulfide solvent, but the more efficient thermal desorption method became part of the master analytical scheme of the EPA in the 1970s. The methods of purge and trap (typical results for volatiles in water are shown in Figure 5) and DHS (volatiles from solids) were established using EPA protocols by interfacing the methods to GC, GC-MS, and GC-FTIR instrumental systems to provide for fully automated detection and identification capabilities. A commercially available system representing the state of the art in combined GC-MS/FTIR analysis is the Hewlett-Packard IRD/MSD,[4] and an automated sampler/analyzer of drugs and metabolites is illustrated with the high-pressure liquid chromatography (HPLC) Zymate System (Zymark Corporation, Hopkington, MA). A robotic sampling system provides the sample-handling capability for a serum sample with the sequence shown in Figure 6. The final step includes an interface for sample injection into an automated HPLC system. The latter units currently provide a major analytical capability in the biomedical and pharmaceutical fields for drug analysis.[5]

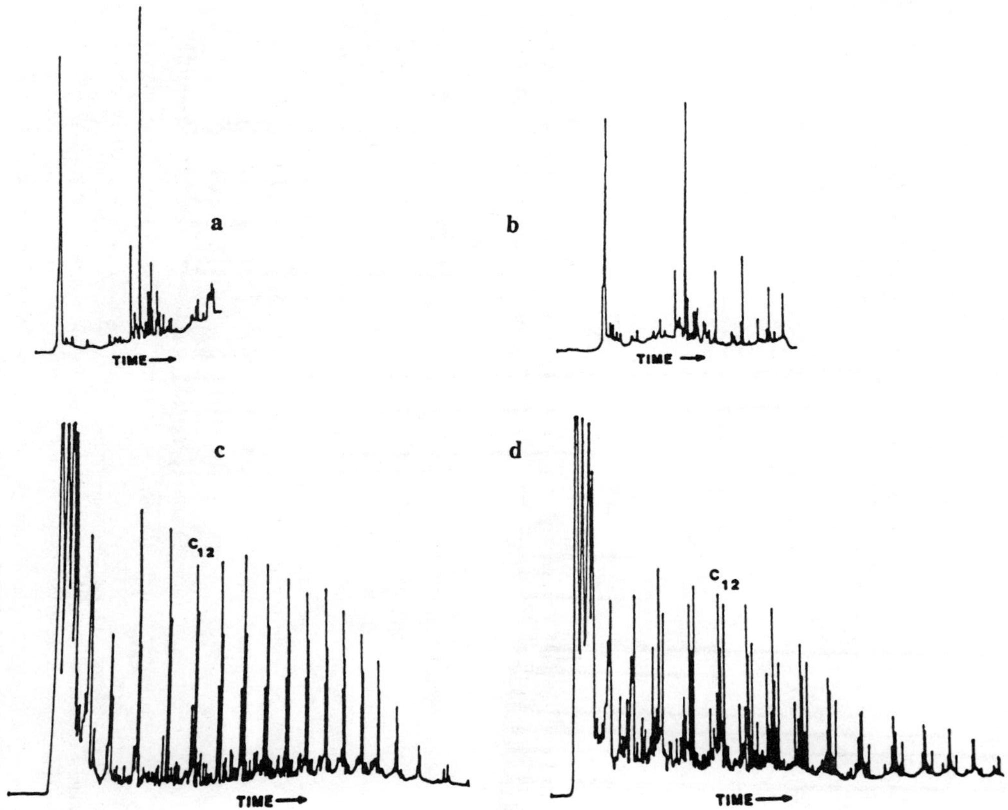

Figure 4. Dynamic headspace (a,b) and pyrolysis (c,d) gas chromatography spectra of a polyethylene-polyvinylacetate copolymer. Atmospheric conditions of each run were helium (a,c) and air (b,d).

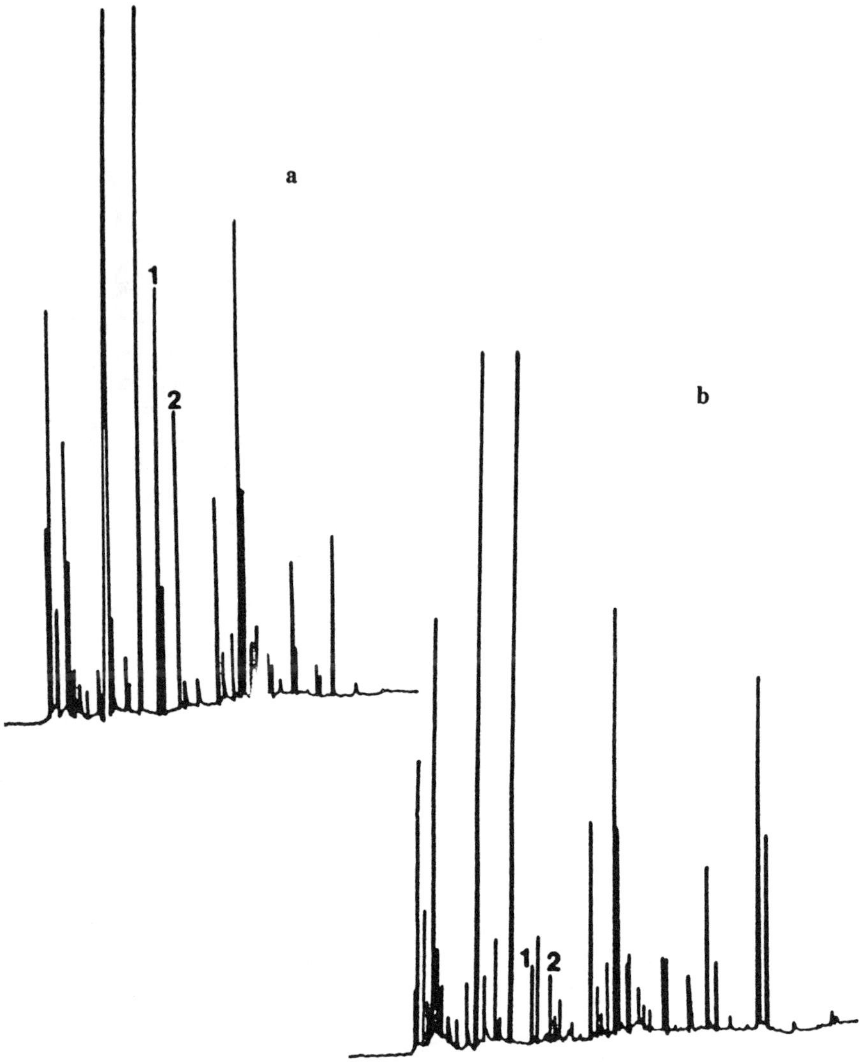

Figure 5. Purge and trap with gas chromatographic analysis of water samples of the ethylbenzene (1) and *meta*-xylene (2) analytes. (a) 100 parts per trillion and (b) 10 parts per trillion of each compound.

INHALATION TOXICOLOGY

Analytical pyrolysis instrumentation and methodology have played significant roles in the development of formulated consumer products that exhibit improved thermal resistance and minimum smoke, flammability, and toxic products formation.[6,7] A general scheme is shown in Figure 7, which outlines

- **ALIQUOT OF 1 ML SERUM SAMPLE**
- **ADD 250 uL INTERNAL STANDARD**
- **ADD EXTRACTION SOLVENT**
- **TUMBLE/MIX**
- **CENTRIFUGE**
- **TRANSFER**
- **RECONSTITUTE IN 250 ML MOBILE FLUID**
- **INJECT INTO HPLC**

Figure 6. Typical sample handling and processing purification procedure of the Zymate robotic sampling system prior to analyte detection by HPLC.

PROBLEM:

- **FORMULATION AND PROCESSING OF MATERIALS TO DECREASE SMOKE, FLAMMABILITY AND TOXICITY**

APPROACH:

- **ESTABLISH ANALYSIS/TEST PROTOCOLS FOR CONSUMER PRODUCTS**
- **DEVELOP ANALYTICAL INSTRUMENTATION/METHODS FOR MATERIALS IN FORMULATED PRODUCTS**
- **ESTABLISH THERMAL DEGRADATION PROFILE**
- **RELATE ANALYTICAL DATA TO TOXICITY EVALUATIONS, BIOASSAYS AND ANIMAL RESPONSE**

RESULT:

- **ANALYTICAL TOXICITY INDEX CORRELATED TO INHALATION TOXICITY TESTS**

Figure 7. National Institute of Standards and Technology (NIST) inhalation toxicology protocol for consumer products.

a general industrial R&D protocol that was established in cooperation with the National Institute of Standards and Technology. These efforts improved formulated products, including microencapsulation technology, to achieve required performance criteria. A comprehensive treatment of inhalation toxicity is provided in a recent book edited by H. Salem.[8]

MICROCAPSULE ANALYSIS

Since natural and synthetic polymers are used in the microencapsulation field as the "shell" containing the "core" (target species and formulation additives), the background analytical capabilities discussed above are directly relevant. Structures of some common shell polymers, e.g., glycogen, cellulose acetate, alginic acid, and polyacrylamide, are shown in Figure 8. In a cooperative research effort with Bend Research Corporation, Bend, OR, analyses of two microencapsulated samples are shown in Figure 9a,b with two different polymers which form shells around the diazinon pesticide core. The analyses were performed with a Pyroprobe (Chemical Data Systems, Oxford, PA) programmable pyrolyzer/concentrator system interfaced to a Finnigan-MAT TSQ 45 mass spectrometer. Total ion chromatograms were obtained for the microencapsulated diazinon samples (Figure 9). The last GC peak in each run contained the volatilized pesticide. Figure 10 presents the electron-impact mass spectrum of the last GC peak scan (#666) in Figure 9a,b and was identified as the diazinon core species by a mass spectral library matching algorithm.

Microgram amounts of microencapsulated materials were also investigated by triple quadrupole daughter ion (MS/MS) MS[9] (Figure 11) with linearly programmed (controlled) sample heating at atmospheric pressure. The Pyroprobe pyrolyzer was directly inserted into the atmospheric pressure chemical ionization (APCI) source of the MS/MS system where soft ionization of the sample occurs by proton attachment from protonated water clusters in the ion source. An investigation was conducted for the active ingredient in the Contac® allergy- and cold-relief consumer product. Figure 12 shows the pyrolysis-APCI mass spectra of the white, pink, and yellow-colored beads in a Contac® capsule. Note that the pink and white beads produce identical mass spectra (Figure 12a,b), while the yellow bead (Figure 12c) lacks the m/z 135 and 152 ions. Identification of a particular mass takes place by selecting a single ion in the mass spectrum and colliding the ion with an argon collision-induced dissociation gas. The ion breaks up into characteristic lower-molecular weight fragments, and these masses are recorded as a daughter ion mass spectrum. The daughter ion mass spectrum of m/z 135 of the pink beads was identical to that of the white beads and is shown in Figure 13a. Figure 13b presents the structure of the vasoconstrictor phenylpropanolamine hydrochloride and is the principle active ingredient by weight in a Contac® capsule. The daughter ion spectrum (Figure 13a) matches that of pure protonated phenylpropanolamine where m/z 43 and 57 can be the $[H_2NC(H){=}CH_2]^+$ and $[HC{=}CH(NH_2)CH_3]^+$ fragments, respectively.

a

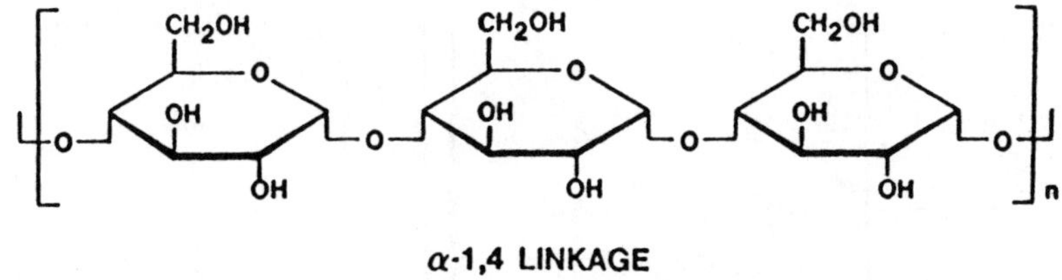

b

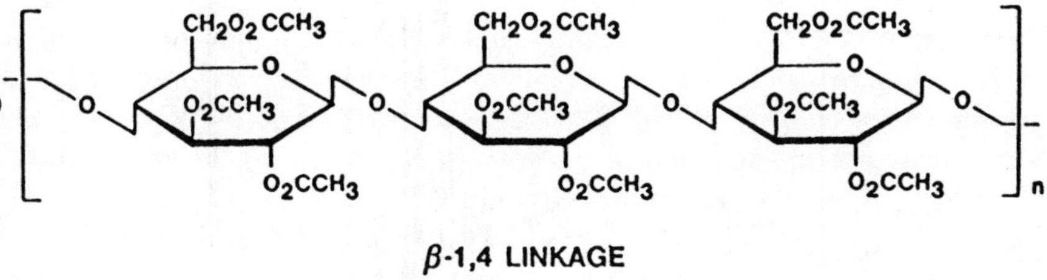

c

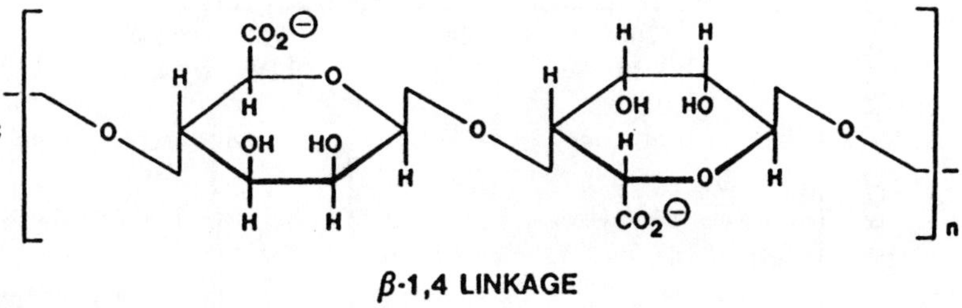

d

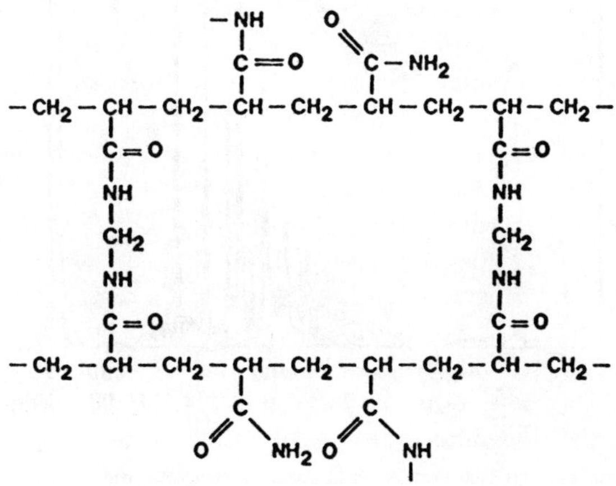

Figure 8. Structures of polymers used as microcapsule shells: (a) glycogen, (b) cellulose acetate, (c) alginic acid, and (d) polyacrylamide.

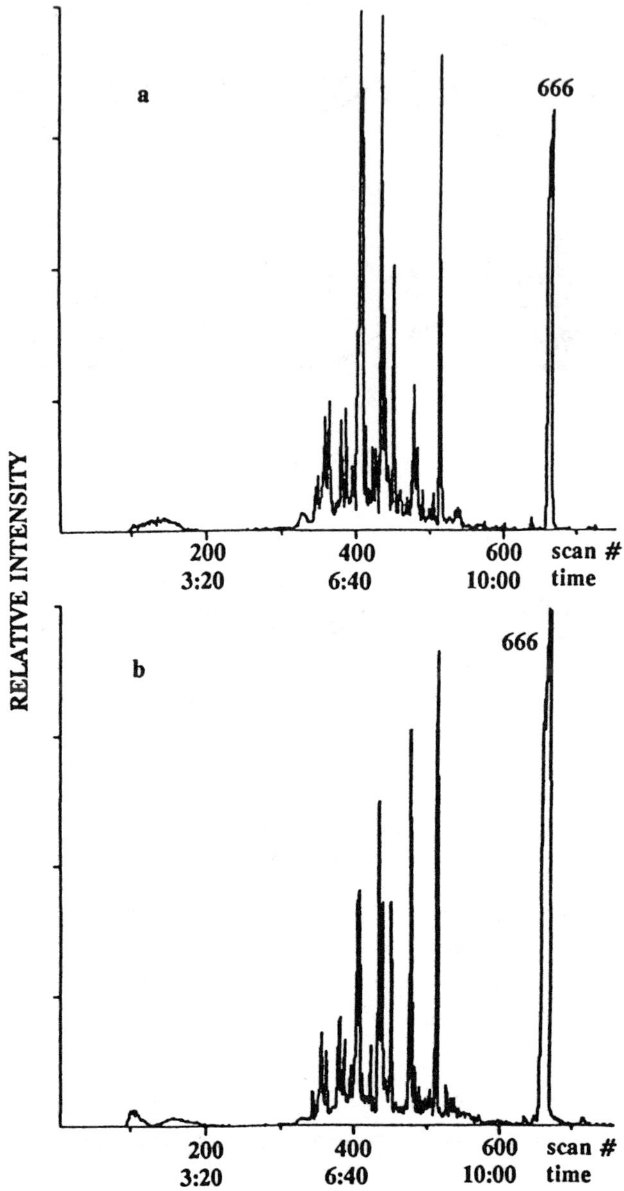

Figure 9. Pyrolysis-gas chromatography-mass spectrometry total ion current of microencapsulated diazinon pesticide. Heating of the solid was accomplished at 240°C/min to a final 700°C temperature. (a) Polymer A shell, (b) Polymer B shell.

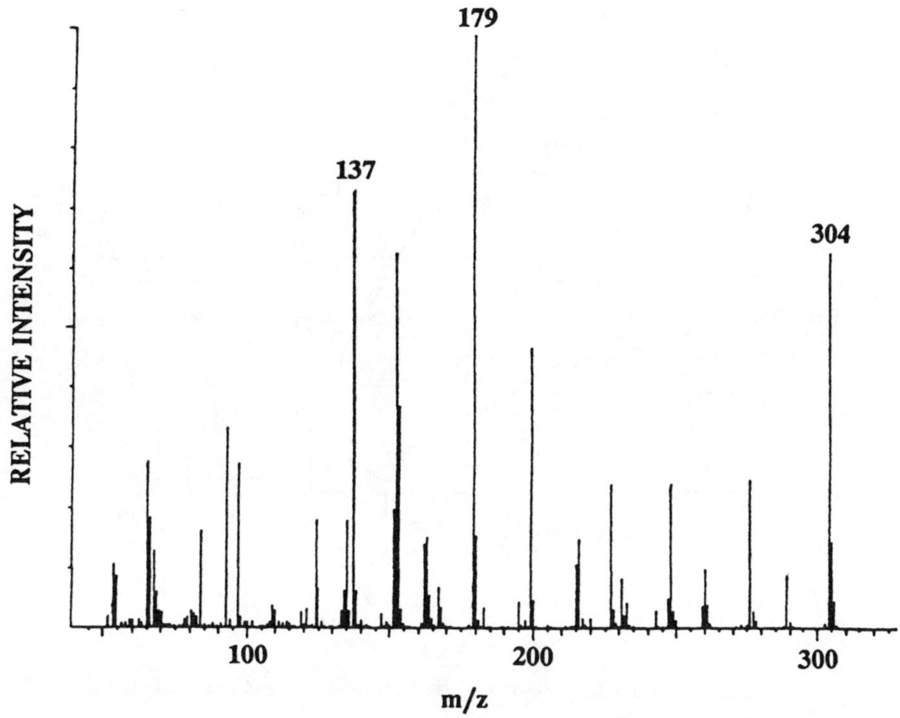

Figure 10. Electron impact mass spectrum of the pesticide diazinon, eluting from the GC column at scan number 666 (Figure 9a,b).

A pyrolysis as well as a direct analysis (DHS) of the No-Pest® Strip consumer product is presented where the controlled release formulation is analyzed with and without thermal treatment (Figure 14). From the strip itself, simple volatilization of the active ingredient, dichlorvos, permits detection and identification directly with this sensitive APCI MS/MS system. Dichlorvos contains two chlorine atoms, and therefore the characteristic isotopic intensity ratio of the m/z 221, 223, and 225 protonated dichlorvos parent ions are observed as 100:66:10, respectively. It appears that by pyrolyzing (heating) the strip, a smaller amount of the pesticide is observed (Figure 14b). These and other relevant illustrations are reported elsewhere.[10,11]

Additionally, the current ultra-trace detection system for proteins/peptides and amino acids in water is best illustrated in reports by Divichi.[12] Using relatively simple capillary electrophoresis equipment with dual lasers which employed the thermal lens effect, analysis of derivatized amino acids was reported at attomole (10^{-18} mol) detection levels. When such units are interfaced to MS or MS/MS systems the final identification step is achieved, as demonstrated by Smith and co-workers.[13]

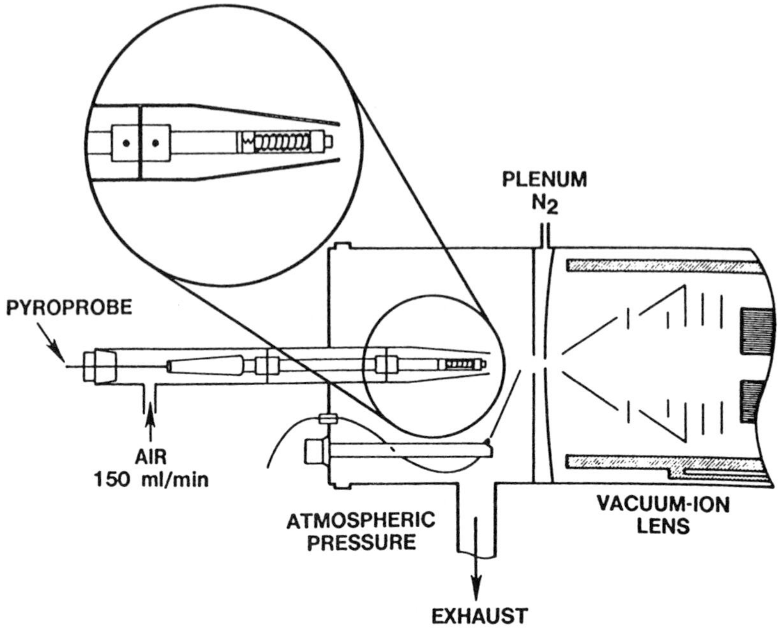

Figure 11. Schematic of a Sciex TAGA 6000 atmospheric pressure chemical ionization triple quadrupole mass spectrometer.

SUPERCRITICAL FLUID TECHNOLOGY

Using supercritical fluid extraction (SFE) and SF chromatography (SFC) instrumentation schematically outlined in Figure 15, a wide range of materials may be extracted, desorbed, or reacted under sub/supercritical fluid conditions.[14,15] This permits the separation of thermally labile compounds from their matrices with minimal thermal treatment.

Isolation and identification of a chemical warfare nerve agent simulant, bis-(ethylhexyl)phosphonate, in a soil matrix were performed using SFE/SFC instrumentation (Computer Chemical Systems, Avondale, PA). These studies record the fate of select compounds under the biodegradation phenomena of soil ecosystems. A control soil sample was extracted first by SFE over time and at a selected extraction time, the supercritical eluate was transferred automatically into the on-line SFC instrumentation. These results were then compared to that obtained with the soil matrix spiked with bis-(ethylhexyl)phosphonate which had undergone the same SFE-SFC operation. Estimates from these data gave detection levels of the target chemical in the low nanogram range for such analyses.

SFE profiles were obtained using FID to monitor the organic compounds extracted from the polymer A shell/diazinon pesticide microencapsulated

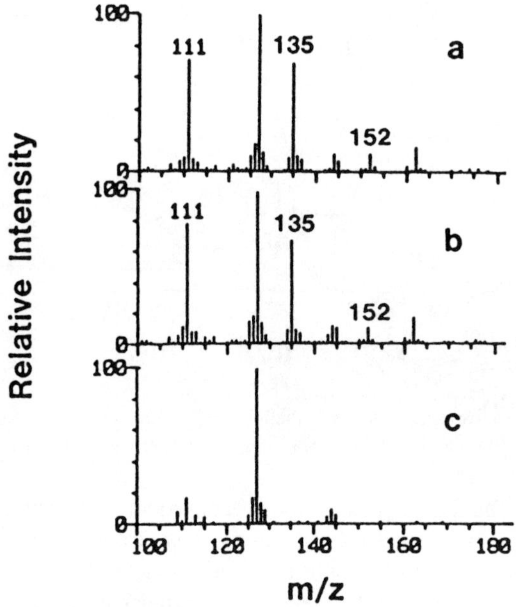

Figure 12. Pyrolysis (500°C)-APCI-mass spectra of each colored bead in the Contac® cold remedy formulation; (a) pink, (b) white, and (c) yellow bead.

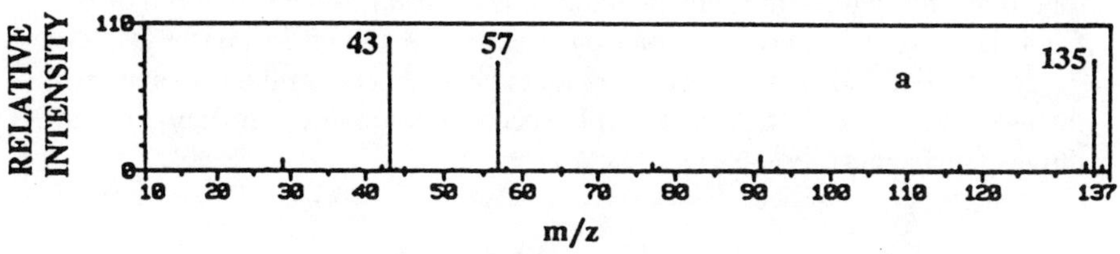

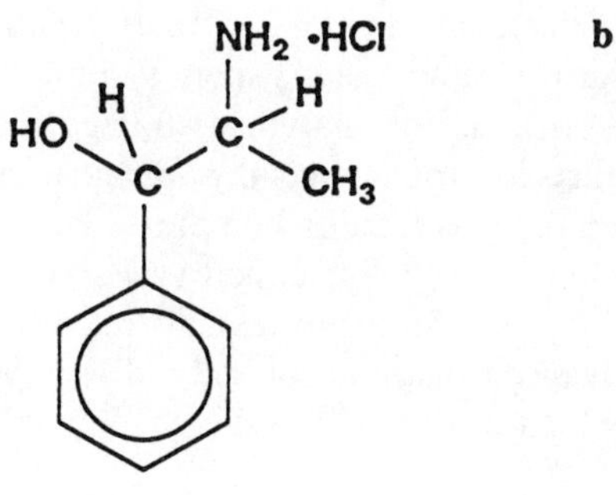

Figure 13. (a) Pyrolysis-APCI-daughter ion mass spectrum of the protonated phenylpropanolamine ion at m/z 135 in Figure 12a,b. (b) Structure of the vasoconstrictor phenylpropanolamine hydrochloride salt.

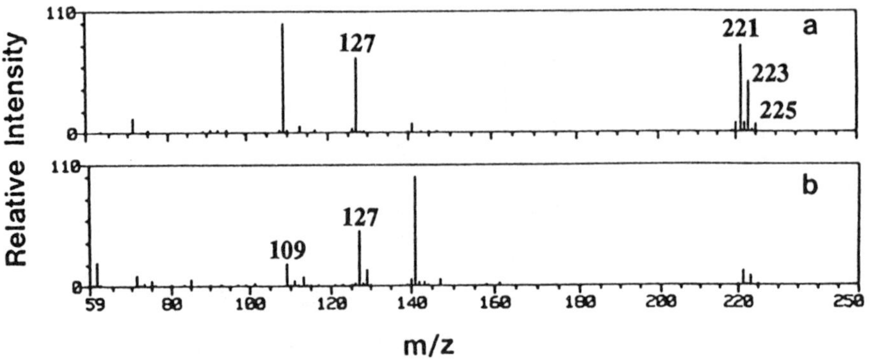

Figure 14. APCI-mass spectra of the (a) volatiles directly from an unadulterated No-Pest® Strip and (b) pyrolyzate from the pyrolysis of a 0.1-mg sliver of the strip product.

sample that was analyzed previously by analytical pyrolysis (Figure 9a). In method development, experimental designs are used to efficiently arrive at optimal settings and sample conditions and to determine a reasonable "response surface" appropriate to the specific experimental configuration. Experimental parameters such as SFE pressure, SFE temperature, and SFC parameters were systematically varied over defined limits. The Box-Behnken experimental design[16] was used for this purpose, and 3 experiments (from a total of 46 needed for the statistical analysis of results where each run differs from the next by a change in one parameter) are shown in Figure 16. The pesticide core, diazinon, is readily extracted and appears as the first and most intense peak. Later peaks in the SFE spectra comprise the formulation additives (surfactants, emulsifiers, etc.).

APPLIED ARTIFICIAL INTELLIGENCE

With the complexity of the problem statements and diversity of materials to sample, detect, and identify at trace levels, analytical chemists have a formidable task. Commercial expert system shell development tools make possible the encoding of analytical strategy and accompanying expertise of experimentalists in various specialty domains. The use of artificial intelligence (AI) has been demonstrated in complex analyses, such as microencapsulated materials, within a prototype expert system network for materials analysis.[17] This technology transfer effort demonstrated how government-oriented goals requiring advanced instrumentation and methods has led to applied AI development which has benefit to the general scientific community.

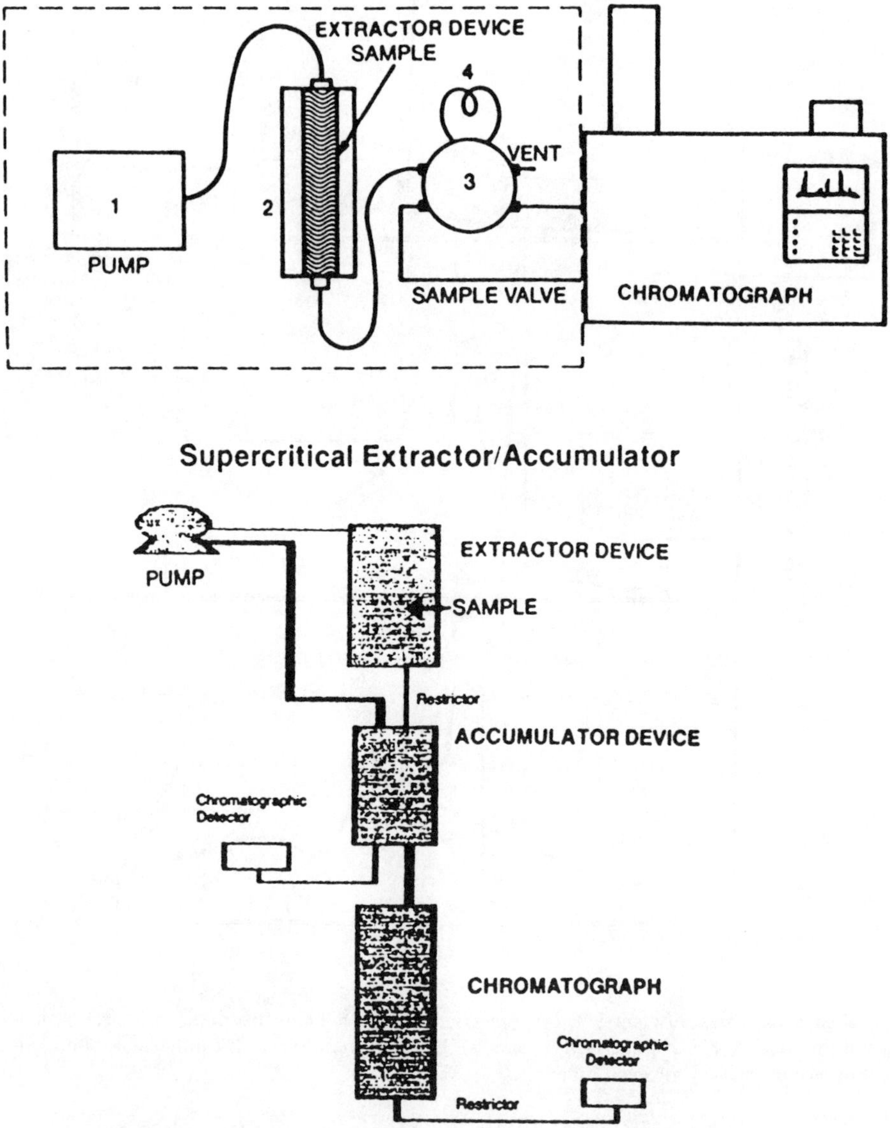

Figure 15. Schematic of a supercritical fluid extraction/chromatography system.

SUMMARY

The analytical summary for this overview has included illustrations of (1) the integrated intelligent instrument (I^3) approach involving analytical

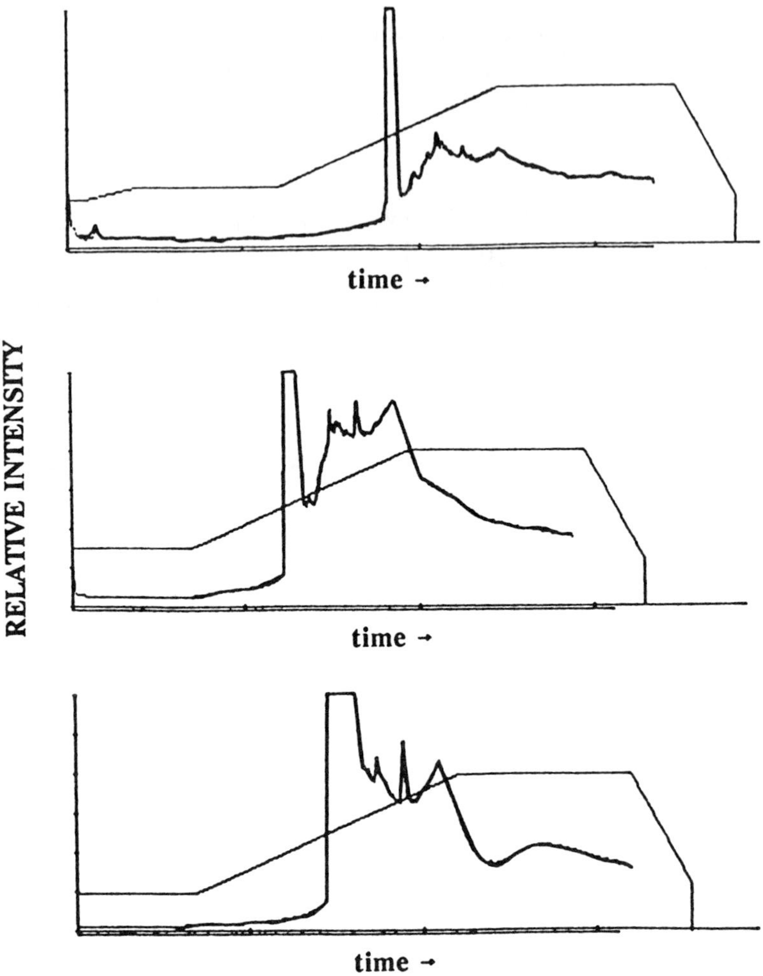

Figure 16. Three different SFE experiments in the extraction of the diazinon pesticide from the polymer A microcapsule shell (Figure 9a). Each spectrum represents a different combination of instrumental/experimental parameters.

strategy, instrumentation, and methods; (2) general complex mixture analyses for the materials and environmental sciences; and (3) applied AI to encode information into an expert systems network.

The philosophical summary may be stated in the following manner: (1) *fact*: analytical toxicologists conduct trace organic analyses with highly automated commercial instrumentation in versatile configurations for wide-ranging chemical/biological species and (2) *challenge*: toxicologists must make proper evaluations of these data, incorporate them into the risk-assessment arena, and integrate the judgmental results into public, political, legal, and moral problem-solving situations.

ACKNOWLEDGMENT

The authors would like to thank Ms. Linda G. Jarvis for the preparation and editing of the manuscript.

REFERENCES

1. *Pyrolysis and GC in Polymer Analysis.* S. A. Liebman and E. J. Levy, Eds., Marcel Dekker, New York (1985).
2. Chemical Data Systems, Oxford, PA 19363, Sample Concentrator Models 310, 320, 330.
3. Saferstein, R.: Pyrolysis in forensic science, in *Pyrolysis and GC in Polymer Analysis,* S. A. Liebman and E. J. Levy, Eds., Marcel Dekker, New York (1985), chap. 7.
4. Hewlett-Packard, Avondale, PA, GC-IRD/MSD Systems for Integrated Fourier Transform Infrared and Mass Spectral Compound Identification.
5. *Instrumental Data for Drug Analysis,* T. Mills and J. C. Roberson, Eds., Elsevier Science Publishers, New York (1987).
6. Liebman, S. A., D. H. Ahlstrom, C. I. Sanders, E. J. Quinn, and C. D. Nauman: *JFF/Combust. Toxicol.* 1:78 (1974).
7. Liebman, S. A., P. J. Duff, K. D. Fickie, M. A. Schroeder, and R. A. Fifer: Degradation Profile of Propellant Systems with Analytical Pyrolysis/Concentrator/GC Technology. *J. Hazardous Mater.* 13:51 (1986).
8. *Inhalation Toxicology,* H. Salem, Ed., Marcel Dekker, New York (1986).
9. Liebman, S. A., A. P. Snyder, J. H., Kremer, D. J. Reutter, M. A. Schroeder, and R. A. Fifer: Time-resolved analytical pyrolysis studies of nitramine decomposition with a triple quadrupole mass spectrometer system, *J. Anal. Appl. Pyrol.* 12:83 (1986).
10. Snyder, A. P., C. W. Cross, S. A. Liebman, G. A. Eiceman and R. A. Yost: Thermal Desorption-Tandem Mass Spectrometry for the Detection and Identification of Formulated Products. *J. Anal. Appl. Pyrol.* 16:191 (1989).
11. Snyder, A. P., J. H. Kremer, S. A. Liebman, M. A. Schroeder, and R. A. Fifer: Characterization of cyclotrimethylenetrinitramine (RDX) by N,H isotope analyses with pyrolysis-atmospheric pressure ionization tandem mass spectrometry. *Org. Mass Spectrom.* 24:15 (1989).
12. Divichi, N. J.: *Chemistry in Britain.* Aug./Sept. (1988), pp. 895–898.
13. Smith, R., J. A. Olivares, N. T. Nguyen, and H. R. Udseth: Capillary zone electrophoresis-mass spectrometry using an electrospray ionization interface. *Anal. Chem.* 60:436 (1988).

14. *Supercritical Fluid Extraction*. M. McHugh and V. Krukonis, Eds., Butterworth, New York (1986).
15. *Chemical Engineering at Supercritical Fluid Conditions*. M. E. Paulaitus, J. M. L. Penninger, R. D. Gray, Jr., and P. Davidson, Eds., Ann Arbor Science, Ann Arbor, MI (1983).
16. Box, G. E. P. and D. W. Behnken, *Technometrics*. 2:455 (1960).
17. *Artificial Intelligence Applications in Analytical Chemistry,* ACS Symp. Ser. No. 306, American Chemical Society, Washington, D.C. 1985. Concerted Organic Analysis and Expert Systems.

Chapter

3

Immunogenicity and Toxicology of Polyethylene Glycol-Modified Enzymes

Friedericka Fuertges and Abraham Abuchowski
ENZON, Inc.
South Plainfield, NJ

Many proteins, peptides, and hormones from a variety of nonhuman sources have potential therapeutic applications. However, these proteins are limited by their short circulating lives and immunogenicity. Repeated administration of the protein often results in the appearance of antibodies which cause increased clearance and may result in severe allergic reactions and anaphylaxis-induced death. It was thought that the use of recombinant human proteins would eliminate these shortcomings. However, while the immunogenicity and antigenicity of recombinant proteins have been somewhat decreased, the problems of a short circulating life still exist.

To circumvent these problems, a number of researchers have developed modification techniques to improve the pharmacokinetic profile and decrease the immunogenicity of proteins. One of the most extensively studied modification techniques is the covalent linkage of monomethoxypolyethylene glycol (PEG) to the amino groups on the surface of the protein. PEG is a linear, hydrophilic, uncharged, flexible polymer which is available in a variety of molecular weights. It is approved by the FDA as a vehicle or base in pharmaceutical preparations. PEG has a very low order of toxicity in oral, parenteral, and epidermal applications (Shaffer and Critchfield, 1947; Smyth et al., 1947; Carpenter and Shaffer, 1952; Smyth et al., 1950).

The modification of proteins with PEG results in an adduct with an increased circulating life and decreased immunogenicity and antigenicity. The

0-8493-8896-1/93/$0.00 + $.50

increase in circulating life is attributed to the masking of receptors associated with clearance. The decreased immunogenicity of the modified proteins is thought to be due to the passive steric hindrance of the antigenic determinants on the protein or to an active effect on the immune system that induces the formation of T suppressor cells specific for the PEG-antigen (Lee et al., 1981). The degree of immunogenicity associated with a PEG-modified protein varies with each protein and the amount of PEG attached to the protein surface (Richter and Akerblom, 1983).

Over 40 proteins have been modified with PEG, and all demonstrate similar alterations in the above mentioned characteristics. Currently there are four PEG-modified enzymes undergoing clinical trials. PEG-L-asparaginase is in phase III trials for the treatment of acute lymphoblastic or lymphocytic leukemia. PEG-adenosine deaminase (ADA) is used in enzyme-replacement therapy for ADA-deficient severe combined immunodeficiency syndrome. PEG-superoxide dismutase (PEG-SOD) is used in the treatment of reperfusion injury associated with kidney transplantation. PEG-uricase is a uricolytic agent used in the treatment of hyperuricemia associated with chemotherapy. PEG-catalase is in preclinical trials and shows potential therapeutic use in trauma, burns, and other hydroxyl radical-mediated injuries. Preclinical research has shown that these compounds have a very low toxicity, as well as reduced immunogenicity and antigenicity.

L-Asparaginase is isolated from either *Erwinia chrysanthemi* or *Escherichia coli*. Both of these preparations are immunogenic and have caused anaphylactic reactions during clinical use. The development of neutralizing antibodies against L-asparaginase has been associated with the rapid clearance of the enzyme and the reduction or loss of enzymatic and antitumor activity.

The modification of L-asparaginase with PEG increases the circulating life in mice from 1 to 13 d (Viau et al., 1986b). PEG-L-asparaginase has a lower order of toxicity than the native enzyme. Both native and modified asparaginases caused weight loss in dogs. However, PEG-L-asparaginase caused no significant toxic effects such as pancreatitis, hyperpnea, urticaria, or anorexia/vomiting (Teske et al., 1988). While anemia has resulted following L-asparaginase administration, anemia has not been observed with the use of PEG-L-asparaginase (Viau et al., 1986b).

Modification of L-asparaginase with PEG markedly reduced the immunogenicity of the enzyme. Animals treated with PEG-L-asparaginase did not demonstrate clinical signs of immunologic toxicosis and did not exhibit anaphylactic responses after repeated drug administration (Viau, 1986b). Mice immunized against the PEG-L-asparaginase did not develop circulating antibodies detectable by immunodiffusion to either the modified or the native enzyme. Rabbits immunized with PEG-L-asparaginase demonstrated decreased antibody titers (Abuchowski et al., 1984).

PEG-modified adenosine deaminase (PEG-ADA) also demonstrates an increased blood circulating life, a very low order of toxicity, and decreased immunogenicity. Modification of the enzyme increased the circulating life in mice from 30 min to 28 h and resulted in a decrease in immunogenicity. Cross reactivity between PEG-ADA and native ADA was not seen in mice immunized with native ADA (Davis et al., 1981).

Like other PEG-modified enzymes, PEG-catalase demonstrates an extended circulating life and decreased immunogenicity and antigenicity as compared to the unmodified enzyme. The adduct had a circulating half-life of 52 h compared to the native enzyme half-life of 2 h. This is true following repeated injections (Abuchowski, 1977).

PEG-catalase demonstrates a low order of toxicity. Following PEG-catalase administration, morphological differences in the spleen consisting of increased extramedullary myelopoiesis with a predominance of erythrocytic precursor cells were noted. This effect was both dose and time dependent. All hematological values, urinalysis results, blood chemistry, and microscopic evaluation were normal (Viau et al., 1986a).

Plasma from mice immunized with catalase tested against the PEG-catalase antigen yielded an antibody titer 50-fold less than that of the same antisera tested against catalase. The route of injection did not appear to affect the antibody titer (Nucci et al., 1986).

PEG-SOD demonstrates characteristics similar to other PEG adducts. PEG-SOD has a circulating life of 8 d in rats, while unmodified SOD is cleared from the circulation in 4 h following the first injection. Toxicity is minimal, consisting of cytoplasmic vacuolation of splenic macrophages in rats. This effect occurred in a dose-dependent incidence and severity. Like PEG-catalase, the parameters measured following PEG-SOD administration were normal (Viau et al., 1986a).

Modification of SOD with PEG results in a reduction in antigenicity and immunogenicity. When PEG-SOD was administered to mice by the i.m. or i.p. route, the titers obtained were less than a tenth of a percent of that of the unmodified enzyme. Antisera to PEG-SOD reacted with higher titers against the corresponding unmodified antigen than the immunizing antigen. The route of administration did not play a major role in the immune response to the modified enzymes (Nucci et al., 1986).

PEG-uricase also demonstrated a very low order of toxicity and reduced immunogenicity. All parameters measured for toxicologic evaluation were normal. Immunological studies of PEG-uricase in mice and rabbits demonstrate a relative lack of immunogenicity and antigenicity in comparison with the native uricase. Administration of native uricase in mice caused severe anaphylactic reactions, while no anaphylaxis was seen in mice receiving PEG-uricase (Savoca et al., 1984).

REFERENCES

Abuchowski, A., G. M. Kazo, C. R. Verhoest, et al.: Cancer therapy with chemically modified enzymes. I. Anti-tumor properties of PEG-asparaginase conjugates. *Cancer Biochem. Biophys.* 7:175 (1984).

Abuchowski, A., J. R. McCoy, N. C. Palczuk, et al.: Effect of covalent attachment of polyethylene glycol on immunogenicity and circulating life of bovine liver catalase. *J. Biol. Chem.* 252:3582 (1977).

Carpenter, C. P. and C. B. Shaffer: A study of the polyethylene glycols as vehicles for intramuscular and subcutaneous injection. *J. Am. Pharm. Assoc.* 41:27 (1952).

Davis, S., A. Abuchowski, Y. K. Park, and F. F. Davis: Alteration of the circulating life and antigenic properties of bovine adenosine deaminase in mice by attachment of polyethylene glycol. *Clin. Exp. Immunol.* 46:649 (1981).

Lee, W. Y., A. H. Sehon, and E. Akerblom: Suppression of reaginic antibodies with modified allergens. IV. Induction of suppressor T cells by conjugates of polyethylene glycol and monomethoxypolyethylene glycol with ovalbumin. *Int. Arch. Allergy Appl. Immunol.* 64:100 (1981).

Nucci, M. L., J. Olejarczyk, and A. Abuchowski: Immunogenicity of polyethylene glycol modified superoxide dismutase and catalase. *J. Free Radicals Biol. Med.* 2:321 (1986).

Richter, A. W. and E. Akerblom: Antibodies against polyethylene glycol produced in animals by immunization with monomethoxypolyethylene glycol modified proteins. *Int. Arch. Allergy Appl. Immunol.* 70:124 (1983).

Savoca, K. V., F. F. Davis, and N. C. Palczuk: Induction of tolerance in mice by uricase and monomethoxypolyethylene glycol modified uricase. *Int. Arch. Allergy Appl. Immunol.* 75:58 (1984).

Shaffer, C. B. and F. H. Critchfield: The absorption and excretion of the solid polyethylene glycols (''carbowax compounds''). *J. Am. Pharm. Assoc.* 36:152 (1947).

Smyth, H. F., C. P. Carpenter, and C. B. Shaffer: The toxicity of high molecular weight polyethylene glycols: chronic oral and parenteral administration. *J. Am. Pharm. Assoc.* 36:157 (1974).

Smyth, H. F., C. P. Carpenter, and C. S. Weil: The toxicology of the polyethylene glycols. *J. Am. Pharm. Assoc.* 39:349 (1950).

Teske, E., G. R. Rutterman, R. van Heerde, and W. Misdorp: L-Asparaginase versus PEG-L-asparaginase in non-Hodgkins lymphoma: a phase III trial in a canine model. Presented at the Second International Conference of Anti-cancer Research, October 11 to 15, 1988, Saronis, Greece.

Viau, A. T., A. Abuchowski, S. Greenspan, and F. F. Davis: Safety evaluation of free radical scavengers PEG-catalase and PEG-superoxide dismutase. *J. Free Radicals Biol. Med.* 2:283 (1986a).

Viau, A. T., A. Abuchowski, J. R. McCoy, et al.: Toxicologic studies of a conjugate of asparaginase and polyethylene glycol in mice, rats, and dogs. *Am. J. Vet. Res.* 47:1398 (1986b).

Chapter

4

Delivery of Drugs to the Brain by Use of a Sustained-Release Polyanhydride Polymer System

Henry Brem, M.D.
Rafael J. Tamargo, M.D.
Alessandro Olivi, M.D.
Departments of Neurosurgery, Ophthalmology, and Oncology
Johns Hopkins University School of Medicine
600 N. Wolfe Street, Baltimore, MD

Despite advances in pharmacology and surgery, the average time of survival for a patient with malignant glioma remains about 1 year.[12] In particular, the successful application of chemotherapeutic agents for treatment of brain tumors has been hampered by two major problems: the limited ability of many drugs to cross the blood-brain barrier, and the serious systemic toxicity when chemotherapeutic drugs are given in large enough quantities to obtain significant concentrations within the brain.

To obviate these limitations, we have developed a sustained-release polymer system that is implanted directly at the site of the brain tumor and thus bypasses the blood-brain barrier and achieves very high local concentrations of drug with minimal systemic toxicity. The rationale for such an approach is based on the observation of Hochberg and Pruitt[11] that 90% of malignant gliomas recur within a 2-cm margin of the original tumor. Therefore, a therapy targeted at the local site of the tumor is most likely to prevent their recurrence.

0-8493-8896-1/93/$0.00 + $.50

Implantation of radioactive seeds, capable of delivering high doses of radiation locally, has been beneficial in terms of survival of patients with brain tumors.[10] Nevertheless, use of this procedure has been limited by a relatively high incidence (41%) of brain and tumor necrosis necessitating frequent reoperations. In an attempt to increase the tumor selectivity by such an approach and introduce standard and newly developed chemotherapeutic agents, we collaborated with Robert Langer, his research group at MIT, and Mark Chasin of Nova Pharmaceutical Corporation, to develop a biodegradable polymer capable of incorporating chemotherapeutic agents and releasing them over prolonged periods of time.[3,20] The polymer, Biodel®, is a polyanhydride, poly[bis(*p*-carboxy-phenoxy propane anhydride) sebacic acid], (PCPP-SA). The polyanhydrides have a number of features that make them particularly suitable for this study.[5-8,15-17] First, their extreme hydrophobicity causes pure surface erosion. Second, the degradation rate of the polymer and therefore the release rate can range from days to years, depending on the formulation. Pure PCPP degrades over a period of years. Adding sebacic acid (SA) shortens the degradation time and therefore the period of release. Thus, a 20:80 PCPP-SA ratio of polymer releases drug for about 3 weeks. A variety of physical forms with different release kinetics are available, such as sheets, rods, microspheres, or wafers. The material can be modified to be rigid or very flexible. Incorporation of drugs is simple, requires no solvents, and is accomplished at low temperatures.

Before testing the use of the polymers in the human brain, we undertook a series of studies in a variety of animals to determine the safety of implanting the polymer, biocompatibility with the brain tissue, the availability of any drug incorporated into the polymer, and the effectiveness of the polymers with incorporated drugs to control experimental intracranial brain tumors. These studies are described below.

BIOCOMPATIBILITY IN RABBIT BRAIN

To determine biocompatibility in the rabbit brain, 20 rabbits were subjected to bilateral craniotomy. A 50:50 solution of PCPP-SA was administered to 20 hemispheres and 20 hemispheres received gelfoam implants. Gelfoam is a standard neurosurgical hemostatic material which is biodegradable and is routinely left in the human brain at the time of surgery. The rabbits were evaluated twice daily for signs of neurologic or systemic toxicity. The groups were sacrificed sequentially between postoperative day 1 and day 60. We found no signs of neurologic or systemic toxicity; all animals survived to the scheduled sacrifice date and the inflammatory reaction around the PCPP-SA was comparable to that around gelfoam.[2]

BIOCOMPATIBILITY IN RATS

A similar study of biocompatibility of PCPP-SA in the rat brain was carried out, except that 42 hemispheres received PCPP:SA (20:80 solution), 35 hemispheres received gelfoam, and 35 hemispheres received Surgicel® (oxidized cellulose, which is another standard neurosurgical implant). We found no signs of neurologic or systemic toxicity; all animals survived to the scheduled sacrifice date, the inflammatory reaction around the PCPP-SA was comparable to that around Surgicel®, and the PCPP-SA was degraded completely within 4 weeks.[19] The polymers were also implanted subcutaneously into rats by Laurencin et al.[13] and the animals showed no evidence of toxicity nor was there histomorphological evidence of systemic or localized toxicity associated with the implant material or its components.

BRAIN DISTRIBUTION OF BCNU IN POLYMERS

Once the biocompatibility of the polymer alone was established we were in a position to examine the kinetics of release of a chemotherapeutic agent incorporated into the polymer. We chose to study the nitrosourea 1,3-bis(2-chloroethyl)-1-nitrosourea (BCNU). We incorporated tritiated BCNU into the biodegradable polymers to investigate its distribution over time within the brain. The autoradiography experiment showed a much more extended distribution as compared to direct injections. Drug concentrations two standard deviations above background were detectable 21 d after implantation of BCNU-Biodel®. By contrast, such levels of drug activity were no longer detectable 48 h after direct injection. We also determined that the concentration adjacent to the polymer was 6.5 m*M* and, even at a distance of 10 mm, the concentration was 200 μ*M*. These concentrations are much higher than achieved with standard intravenous administration.[5,9]

BCNU-POLYMER EFFICACY IN THE BRAIN

To test whether or not the BCNU released from the PCPP-SA polymer was biologically active, we explored the efficacy of BCNU-impregnated polymers in the rat brain. We studied 96 rats and divided them into 6 groups performing 9L glioma or sham implantations on the first day. On the fourth postoperative day, BCNU polymer or empty polymer was implanted at the tumor site and intraperitoneal injection of either BCNU or ethanol and normal saline was administered, according to the protocol as shown in Table 1.

We found that with empty polymer, i.e., the untreated controls, the average survival was 11 d, with intraperitoneal BCNU the average survival

TABLE 1
Effect of BCNU Polymer on Survival of Rats Implanted with Gliosarcoma

		Brain implant		Intraperitoneal		
Group	Implantation with 9L glioma	Empty polymer	BCNU polymer	BCNU	Ethanol & normal saline	Mean survival (days)
1	−	+	−	−	+	(125)
2	−	+	−	+	−	63
3	−	−	+	−	−	93
4	+	+	−	+	−	27
5	+	−	+	−	+	62
6	+	+	−	−	+	11

was 27 d, and with intracranial BCNU and PCPP-SA the average survival was 62 d. In addition, two of the nine rats with intracranial BCNU in PCPP-SA survived to the time of planned sacrifice, 120 d, with no evidence of tumor, whereas none of the rats with empty polymers or intraperitoneal BCNU injections survived beyond 58 d.[20]

EVALUATION OF POLYMERS IN MONKEYS

Before initiating clinical studies in humans, we examined the effects of empty and drug-impregnated polymers in 15 *Macaca fascicularis* monkeys. All monkeys underwent a right craniotomy. Five then received an intracerebral implantation of gelfoam, five received Biodel®, and the remaining five were implanted with BCNU-Biodel® polymer. The monkeys were evaluated twice daily for signs of neurologic or systemic toxicity. Blood chemistries and hematology tests were obtained periodically, as were CT scans and magnetic resonance images. In each group, three monkeys were sacrificed early, between days 16 and 18, and two monkeys were sacrificed later, on days 71 and 81. Each animal received a complete autopsy. We found that only one monkey had a left hemiparesis that was related to the surgical technique of lobectomy. There was no systemic toxicity from the implants. All of the monkeys survived to the scheduled sacrifice day. The blood tests were normal throughout. The CT and MRI scans localized the degrading polymers and showed minimal reaction to them. Autopsies were remarkable only for localized reaction around the polymer.[4]

In summary, we had shown that polymeric carriers are biocompatible in the brain of rats, rabbits, and monkeys; that polymeric carriers can release biologically active BCNU in a gradual controlled fashion; and that tumor growth was inhibited more effectively by interstitial chemotherapy than by systemic therapy. We therefore began our clinical studies.

CLINICAL TRIALS

A phase I clinical study was started to investigate the safety of the BCNU Biodel® (PCPP-SA 20:80) implanted at the site of tumor resection.[3] This clinical study was based on the known tendency of malignant brain tumors to recur locally[11] and the fact that BCNU given systemically has a half-life of only 12 min.[18] In this study, the BCNU is protected from degradation within the anhydrous polymer. Therefore, after the tumor was extirpated we covered the surgical cavity with BCNU Biodel® polymer so that the target tumor cells at the time of the resection would be exposed directly to the drug. This study included patients who had previous surgery for grade III or IV gliomas confirmed by histology and previous definitive external beam radiation. The patients had to be 21 years or older with a Karnofsky reading of 60 or higher. The tumor diameter had to be 1.5 cm or greater and the patient had to be capable of executing written informed consent. Patients excluded from this study had at least one of the following characteristics: more than one focus of tumor; less than 100,000 platelets; less than 4,000 leukocytes; a known anaphylactic reaction to nitrosoureas; significant renal or hepatic disease; pregnancy; or a concomitant life-threatening disease.

In this clinical trial, which was sponsored by the Nova Pharmaceutical Corporation, 21 patients were studied at five institutions. Five patients were entered with a 1.93% BCNU loading dose and a surface concentration of 25 μg per square millimeter. Once we determined that this was a safe dose, we enrolled an additional five patients with 3.85% BCNU loading and a surface concentration of 50 μg per square millimeter. Another 11 patients were enrolled with 6.37% BCNU loading in an 82.5-μg per square milliliter concentration 4 months later. The results of this study are currently being analyzed and will be presented in a separate report.

In the eight patients investigated at Johns Hopkins Hospital, we found that the polymers were well tolerated and did not cause any detectable toxicity. We were surprised to find that 2 to 3 months after implantation, a contrast CT scan showed a ring enhancement around the polymers with mild mass effect that gradually resolved. Three of our eight patients are alive 1 year after implantation.

To determine the effectiveness of this treatment approach, Nova Pharmaceutical Corporation is sponsoring a multi-institutional prospective study in the form of a randomized, placebo-controlled, double-blinded trial. Patients who have failed conventional treatment will either receive Biodel® with BCNU or Biodel® without chemotherapeutic drug. This clinical study should enable us to determine the effectiveness of BCNU-impregnated polymers in treating recurrent malignant brain tumors.

There are numerous other applications of the sustained-release polymers for drug delivery in the brain. At present we are investigating the efficacy of release of steroids to treat brain edema and angiogenesis inhibitors to control the biological activity of brain tumors. We hope that this new approach to treating brain tumors will allow us to maximize the benefit of standard chemotherapeutic agents, newly developed drugs, and biological response modifiers while minimizing the systemic toxicities associated with these treatments.

ACKNOWLEDGMENTS

We gratefully acknowledge the assistance of Pamela Talalay, Ph.D.; O. Michael Colvin, M.D.; Kam Leong, Ph.D.; Stuart Grossman, M.D.; Eileen Bohan, R.N.; Donlin M. Long, M.D., Ph.D.; Hyo Ahn, M.D.; Carla Reinhard, MHS; and Michael Pinn, B.A. of Johns Hopkins Hospital; Robert Griffith, M.D.; James Kenealy, Pharm.D.; and Mark Chasin, Ph.D. of Nova Pharmaceutical Corporation; and Robert Langer, D.Sc., of Massachusetts Institute of Technology.

Supported in part by a grant from Nova Pharmaceutical Corporation, NIH Grant No. NS01058-01, American Cancer Society Grant No. IN-11W, the Andrew W. Mellon Foundation Johns Hopkins University Faculty Development Award, and the Association for Brain Tumor Research Fellowship in Memory of Steven Lowe.

REFERENCES

1. Brem, H., D. Fossett, J. S. Myseros, K. W. Leong, R. J. Tamargo, M. Colvin, and J. Hilton: Sustained Local Release of a Biologically Active Chemotherapeutic Agent by a New Bioerodible Polymer. Congress of Neurological Surgeons, Baltimore, October 24, 1987.
2. Brem, H., A. Kader, J. I. Epstein, R. Tamargo, A. Domb, R. Langer, and K. Leong: Biocompatibility of bioerodible controlled release polymers in the rabbit brain. *Sel. Cancer Ther.* 5(2):55–65 (1989).
3. Brem, H.: Polymers to treat brain tumors. *Biomaterials.* 11:699–701 (1990).
4. Brem, H., R. J. Tamargo, M. L. Pinn, and M. Chasin: *Biocompatibility of a BCNU-Loaded Biodegradable Polymer: A Toxicity Study in Primates.* American Association of Neurological Surgeons, Toronto, p. 389, April 24, 1988.
5. Chasin, M., A. Domb, E. Ron, E. Mathiowitz, K. Leong, C. T. Laurencin, H. Brem, and R. Langer: Polyanhydrides as drug delivery systems, in *Biodegradable Polymers as Drug Delivery Systems,* R. Langer and M. Chasin, Eds., Marcel Dekker, New York, 1990, pp 43–70.
6. Domb, A. and R. Langer: Polyanhydrides. I. Preparation of high molecular weight polyanhydrides. *J. Polymer Sci.* 25:3373–3386 (1987).
7. Domb, A. and R. Langer: Polyanhydrides: stability and novel composition. *Makromol. Chem. Macromol. Symp.* 19:189–200 (1988).

8. Domb, A., E. Ron, and R. Langer: Polyanhydrides. II. One step polymerization using phosgene or diphosgene as coupling agents. *Macromolecules.* 21:1925–1929 (1988).
9. Grossman, S. A., C. S. Reinhard, H. Brem, R. Brundrette, M. Chasin, R. J. Tamargo, and O. M. Colvin: The intracerebral delivery of BCNU with surgically implanted bioerodible polymers: a quantitative autoradiographic study. *Proc. Am. Soc. Clin. Oncol.* 7:84 (1988).
10. Gutin, P. H., T. L. Phillips, W. M. Wara, et al.: Brachytherapy of recurrent malignant brain tumors with removable high-activity iodine-125 sources. *J. Neurosurg.* 60:61 (1984).
11. Hochberg, F. H. and A. Pruitt: Assumptions in the radiotherapy of glioblastoma. *Neurology.* 30:907 (1980).
12. Kornblith, P. L. and M. D. Walker: Chemotherapy for malignant gliomas. *J. Neurosurg.* 68:1–17 (1988).
13. Laurencin, C. T., A. J. Domb, C. D. Morris, V. I. Brown, M. Chasin, R. F. McConnell, and R. Langer: High dosage administration of polyanhydrides *in vivo. Symp. Control. R.L. Bioact. Mater.* 14:140–141 (1987).
14. Leong, K. W., B. C. Brott, and R. Langer: Bioerodible polyanhydrides as drug-carrier matrices. I. Characterization, degradation and release characteristics. *J. Biomed. Mater. Res.* 19:941–955 (1985).
15. Leong, K. W., P. D'Amore, M. Marletta, and R. Langer: Bioerodible polyanhydrides as drug-carrier matrices. II. Biocompatibility and chemical reactivity. *J. Biomed. Mater. Res.* 20:51–64 (1986).
16. Leong, K. W., J. Kost, E. Mathiowitz, and R. Langer: Polyanhydrides for controlled release of bioactive agents. *Biomaterials.* 7:364–371 (1986).
17. Leong, K. W., V. Simonte, and R. Langer: Synthesis of polyanhydrides: melt-polycondensation, dehydrochlorination, and dehydrative coupling. *Macromolecules.* 20:705–712 (1987).
18. Loo, T. L., R. L. Dion, R. L. Dixon, and D. P. Rall: The antitumor agent 1,3-bis(2-chloroethyl)-1-nitrosourea. *J. Pharm. Sci.* 55:492–497 (1966).
19. Tamargo, R. J., J. Epstein, C. S. Reinhard, M. Chasin, and H. Brem: Brain biocompatibility of a biodegradable, controlled-release polymer in rats. *J. Biomed. Mater. Res.* 23:253–266 (1989).
20. Tamargo, R. J., J. I. Epstein, M. B. Yang, M. L. Pinn, M. Chasin, and H. Brem: Gliosarcoma: Controlled Release Polymers for Local Therapy. American Association of Neurological Surgeons, Washington, D.C., April 2 to 6, 1989.

Chapter

5

Blue Red Cells: The Association with Liposomes and Red Cells of LacZ Fusion Polypeptides Containing the Membrane-Anchoring Sequence from Cytochrome B_5

S. K. George, Y.-H. Xu, and G. M. Ihler
Department of Medical Biochemistry and Genetics
Texas A&M College of Medicine
College Station, TX

SUMMARY

DNA corresponding to the membrane-anchoring hydrophobic peptide (HP) of cytochrome b_5 was synthesized and cloned into pUC18. In-frame fusions with *lac*Z, the gene encoding *Escherichia coli* β-galactosidase, were constructed, locating the DNA for HP, in one construct (pSKG2), at the *Eco*RI site near the end of the gene and, in another construct (pUR278:HP), at a multiple cloning site at the end of the gene. In the latter case, but not the former, the fusion protein was enzymatically active. Both fusion proteins associated with the bacterial membrane *in vivo* and with liposomes *in vitro*, as does cytochrome b_5 itself. Use of fusion proteins bearing HP enables liposomes bearing any desired protein on their surface to be readily prepared.

0-8493-8896-1/93/$0.00 + $.50

Both fusion proteins and cytochrome b_5 also associated with red cells. Relocating the HP domain to the amino terminus of the LacZ gene did not affect the ability of the fusion protein to associate with liposomes and red cells. Construction of a tripartite fusion protein, in which calmodulin was located at the C terminus of the pUR278:HP construct so that HP was embedded between two large polypeptides, resulted in a fusion protein having both β-galactosidase and calmodulin activities. It spontaneously associated both with liposomes and red cells so that both the amino and carboxyl termini of the protein remain on the same side of the bilayer. Fusion proteins bearing HP containing single-site mutations, including one having a substitution of a polar for a hydrophobic amino acid within HP, associated with membranes.

Proteins able to spontaneously associate with liposomes or red cells might have interesting properties such as increased circulatory lifetimes or unusual immunological properties. Likewise, many similar applications involving interactions with nucleated cells are foreseen and may also provide an avenue for the development of novel therapeutic agents.

INTRODUCTION

The use of red cells as carriers for therapeutic agents was originally proposed by Ihler et al.,[1] who suggested their use as a delivery vehicle in lysosomal storage diseases involving macrophages, particularly in Gaucher's disease. It was shown that glucocerebrosidase (and other enzymes) could be loaded into red cells during hypotonic hemolysis and entrapped there by resealing the holes. It was anticipated that the red cells loaded with glucocerebrosidase might be taken up by Gaucher macrophages, which contain massive amounts of undegraded glucocerebroside derived largely from ingested red and white cells. However, preliminary trials in advanced Gaucher disease were not encouraging.[2] Although it is still possible this procedure could have some prophylactic efficacy, currently it appears that glucocerebrosidase bearing mannose residues, which is transported into target cells, can effectively reduce the consequences of Gaucher's disease.[3] Effective delivery of drugs to macrophage, such as pentamidine for visceral leishmaniasis, has been demonstrated using red cell ghosts.[4] If used as a one-time targeting and delivery vehicle to macrophage, it would not be necessary for the red cells to survive long in the circulation, but prolonged delivery could be achieved if the red cells continued to circulate.

Carefully resealed red cells can circulate for considerable periods of time, and it became evident that the resealed red cells could have a circulatory half-life comparable to that of similarly handled normal red cells, if appropriate resealing procedures which preserved the intracellular contents and the energy state of the cell were used.[5-8] This finding led to the idea of using circulating resealed red cells for long-term delivery of drugs which could escape slowly.[9]

A serious problem, however, is the difficulty of controlling the rate of exit; for example, cells containing gentamicin circulated with a half-life of 22 d, but as gentamicin is a polar compound and does not escape readily, the cells did not maintain effective plasma concentrations.[10] A particularly interesting kinetic aspect of the red cell system is its potential for delivering a constant amount of drug per unit time. Drugs released when senescent red cells are destroyed by phagocytosis would be expected to escape at a rate of about 1% per day, with a total therapeutic span of about 120 d. Intralysosomal release could be readily accomplished by linking a lipid-soluble drug to a hydrophilic or high-molecular weight, nontransportable compound by a chemical linkage susceptible to cleavage by acidic pH or lysosomal enzymes.

The use of circulating carrier red cells containing loaded enzymes to degrade unwanted substances capable of being transported into the red cell was also proposed.[11] Uses which require the protein to act extracellularly, however, leave open the question of how the protein would escape from the red cell. An early application was the use of red cells containing entrapped asparaginase. Initially the observed dramatic lowering of plasma levels of asparagine was attributed to release of the enzyme by low levels of intravascular hemolysis.[12] Subsequently the kinetics of entry of asparagine into asparaginase-containing red cells has been reinvestigated.[13]

Therapeutic uses for proteins, including enzymes and polypeptides, is well established despite some persistent problems that greatly complicate their use. A principal problem is the relatively short half-life of proteins in the blood, as exemplified by the requirement for daily insulin replacement in diabetes. Considerable effort has been expended with reasonable success in devising various procedures for extending the half-life of exogenous proteins. Immunological problems associated with the repeated administration of foreign proteins can now in principle be avoided by the use of proteins derived from cloned genes of human origin. But it is still difficult to employ proteins therapeutically.

Carrier red cells with proteins or enzymes located on the external surface of the cell would potentially be very useful, since the protein or enzyme would be exposed to circulating substances. Substrates unable to enter the red cell or which enter slowly could be degraded by enzymes with their active sites located on the outside of the red cells. Antibodies carried on the surface of red cells might have novel immunological properties. For this approach to be successful, some procedure for inserting proteins into red cell membranes or for binding or linking them to the red cell surface would be needed.

MEMBRANE INSERTION SIGNALS

A great deal is known about the mechanism by which proteins are translocated across membranes and intracellularly targeted to their appropriate locations by special sequences in the protein or the protein precursor.[14] Amino terminal sequences used for transport through the endoplasmic reticulum (ER) or for export in bacteria contain a basic amino terminal region followed by a continuous stretch of hydrophobic amino acids. In the mammalian ER, the signal sequence binds to the ribonucleoprotein signal-recognition particle (SRP), which in turn binds to an SRP receptor protein in the ER. Release of SRP allows the signal sequence to bind to an ER protein, the signal sequence receptor. Translocation of the protein across the membrane then occurs. Translocation appears to be facilitated by or to require an unfolded protein, such as a nascent protein or a protein which is prevented from folding by interaction with another protein.

An important finding in these studies was that nearly any protein bearing the signal sequence could be translocated. This is true of other membrane-localization signals as well. Mitochondrial and chloroplast proteins bear amino terminal sequences which permit the transfer of other proteins attached to the sequence into mitochondria or chloroplasts. Peroxisomal proteins apparently have carboxyl terminal sequences. Nuclear proteins bear lysine-rich internal signals required for protein import through nuclear pores. These signals, along with other sequences that also affect targeting, such as stop-transfer sequences or glycosylation sites (to which are attached carbohydrates, such as mannose-6-phosphate, which are recognized by receptors involved in sorting) determine the intracellular location of the protein or control its secretion from the cell.

Most studies involving membrane localization are concerned with determining how intracellular porteins are targeted to the appropriate location. Here, however, we are concerned with the question of how proteins could be targeted to cell membranes from the outside. The immediate impetus for our studies was the finding that the *lamB* protein, the receptor for bacteriophage lambda, was functional in liposomal membranes and permitted the entry of bacteriophage DNA into the interior of unilamellar liposomes. We speculated that if *lamB* could be inserted into cellular membranes it would be possible to directly inject cloned DNA into cells.[15,16] Many membrane proteins such as *lamB* have been reconstituted into liposomes in a functional configuration by forming the membrane around the protein at the time the liposome was created. In general, however, the proteins will not spontaneously insert into liposomes once the membrane has formed.

CYTOCHROME B_5

Unlike most other membrane proteins, cytochrome b_5 has been reported to spontaneously associate with liposomes. This protein contains two domains, a hydrophilic domain containing iron and a porphyrin ring which is involved in electron transport, and a hydrophobic domain which is entirely responsible for anchoring the protein in the membrane. These two domains are linked by a trypsin-sensitive linker segment. When cytochrome b_5 associates spontaneously with small PC liposomes most of the protein can exchange readily, and under equilibrium conditions it prefers small unilamellar liposomes (SUV) by a factor of 200 over large unilamellar liposomes (LUV).[17] This could indicate that the spatial orientation of lipids in a membrane with a high radius of curvature permits the hydrophobic domain of cytochrome b_5 to insert into the membrane more readily.

Under other circumstances, however, cytochrome b_5 can insert in a nontransferable configuration, such as with dimyristyl PC liposomes or microsomes. Also, a certain fraction (about 10 to 20%) inserts into large PC liposomes in the nontransferable configuration. The molecular differences between the transferable and nontransferable configurations are not known. Both the hydrophilic domain and the carboxyl terminus of the hydrophobic domain of cytochrome b_5 are located on the outside of the PC liposomes so that no charged residues are available to lock the protein in the membrane. This may perhaps account for the exchangeability of the protein from large to small liposomes. If the carboxyl terminus of the protein were to cross the membrane under certain circumstances, it might act as a stop-transfer sequence and lock the protein in the membrane. In this case, the protein might be much less transferable. Whether or not this is the explanation for the observed instances of nontransferable cytochrome b_5 is not known.

POTENTIAL USE OF THE HYDROPHOBIC DOMAIN OF CYTOCHROME B_5 IN FUSION PROTEINS

By synthesis of DNA corresponding to the hydrophobic domain of cytochrome b_5 (referred to here as *hyp*) and fusion to DNA coding for other proteins, a fusion protein having the property of spontaneous insertion into liposomes could be constructed. We believe that this procedure may be of considerable benefit for preparing liposomes bearing specific proteins on their surfaces. For example, liposomes bearing exposed polypeptide hormones might target to specific receptors on certain cell types. However, liposomes do not survive for extended periods of time in the circulation, so that their use in extending the circulatory life span of a protein is limited. So, we have also conducted studies to determine if this fusion protein can associate with or insert into the membranes of red cells, with a view to preparing a long-lived (120 d maximum) carrier.

SYNTHESIS OF *hyp*

von Bodman et al.[18] synthesized DNA which would code for the entire rat hepatic cytochrome b_5 and expressed the gene in *E. coli*, where it was

1. 5′GCGGCGGCGGATCCACTGTCCAAACCGATGGAAACCCTGATCACCACCGTTAACTCCAACTCC3′

2. 5′CGGGATAACCCAGTTGGTCCACCAAGAGGAGTTGGAGTTAACGGT 3′

3. 5′TCTTGGTGGACCAACTGGGTTATCCCGGCTATCTCCGCTCTGATCGTTGCTCTGATGTACCGT 3′

4. 5′ATATATAGTCGACTCCTAATCGTCAGCAGCCATGTACAGACGGTACATCAGAGCAACGATCAGAGCGGAGATAGC 3′

Figure 1A. Four chemically synthesized molecules.

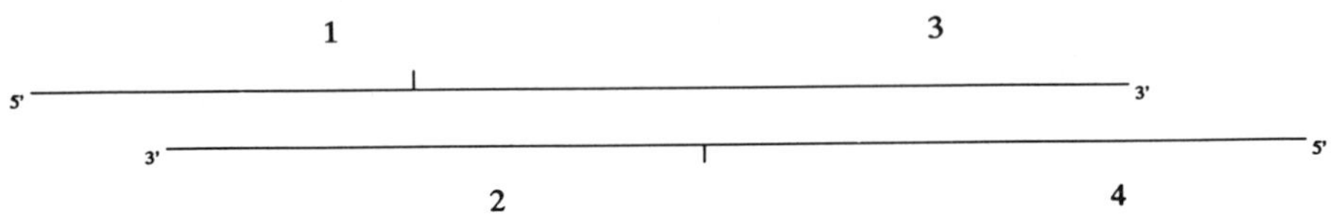

Figure 1B. Duplex DNA prepared by hybridization and ligation of nicks.

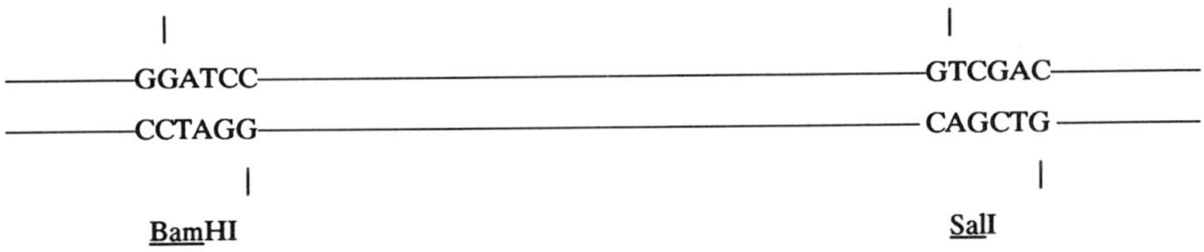

Figure 1C. Duplex DNA after primer extension showing location of *Bam*HI and *Sal*I sites.

found in the membrane fraction. We have synthesized DNA corresponding only to the HP of rabbit cytochrome b_5. Four DNA molecules (Figure 1a) were chemically synthesized having an appropriate coding sequence using codons which are frequently used in *E. coli*.[19] When annealed, a duplex DNA resulted (Figure 1b), having two single-strand breaks which were repaired with T4 DNA ligase and two single-stranded ends which were filled by primer extension with Klenow fragment DNA polymerase. This DNA had a *Bam*H1 site at one end and a *Sal*I site at the other (Figure 1c).

The DNA was first cloned into the multiple cloning site of the expression of vector pUC19 (Bethesda Research Laboratories), where it is inserted in the reverse orientation for transcription. This was done to ensure that the DNA could be cloned in the absence of any possible adverse effect on the viability of the transformant which might be associated with translation of the *hyp* sequence. The orientation was then reversed by cloning it into plasmid pUC18, which is identical except that it has the reverse sequence of restriction sites. This plasmid can serve either as a source of *hyp* DNA for cloning into other plasmids to construct fusion proteins, or alternately, DNA corresponding to other proteins may be cloned into pUC18hyp.

RECOMBINANT POLYPEPTIDES CONTAINING HP

laZ-hyp Fusions

A *Hind*III-*Bam*HI fragment carrying part of the *trp* operon and the *lac* operon genes *Z, Y,* and *A* was excised from pMC81 and cloned into the *Hind*III and *BAM*HI sites downstream from *lacPO* in a derivative of pBH20 which had the *Eco*R1 site removed by cleavage, primer extension, and blunt end ligation. This plasmid is called pSKG1. The *Bam*HI-*Sal*I fragment corresponding to *hyp* was obtained from pUC18hyp. After filling in the single-stranded ends of *hyp* with Klenow fragment DNA polymerase, *Eco*RI linkers were attached by blunt end ligation. Using pSKG1, this DNA was then inserted into the *Eco*RI site located near the C terminus in *lac*Z. This creates a fusion protein bearing 1006 β-galactosidase amino acids (lacking the C terminal 18 amino acids) and 48 new amino acids, including the 45 amino acids of the HP of cytochrome b_5.

This fusion protein lacks the C terminus of β-galactosidase which is required for normal enzymatic activity. When the DNA is transformed into a strain containing the ω-fragment of β-galactosidase, blue colonies result on X-gal, indicating that there is at least a low level of ω-galactosidase activity in these strains. Probably the enzyme is protected or stabilized by the ω fragment, since, when the same DNA is transformed into strains lacking the ω fragment, the colonies are all white.

Enzymatically active fusion proteins have been constructed using the pUR cloning vectors developed by Ruther and Muller-Hill.[20] pUR278 contains a multiple cloning site with unique cloning sites (*Bam*HI, *Sal*I, *Pst*I or *Xab*I, and *Hind*III) located immediately after the AAA codon corresponding to the C terminal lysine. The *Bam*HI-*Sal*I site of pUR278 permits the in-frame cloning of the *hyp* fragment originally cloned into pUC18.

Relocation of HP to the Amino Terminal End

Plasmid vector pSKS105 was restricted with *Xma*I, blunt ended by primer extension with polymerase and cleaved with *Hind*III. A *Sma*I/*Hind*III fragment containing the *hyp* nucleotide sequence was excised from pUC19:hyp and ligated in frame with *lac*Z in the modified pSKS105 vector. The resulting plasmid, pSKS105:hyp, codes for an enzymatically active HP:LacZ fusion polypeptide having the HP sequence located near the N terminus. The HP termination codon (UAG) is suppressed by supE44 (in strain MB392) to obtain the full-length fusion polypeptide-translation product.

Site-Specific Mutagenesis of *hyp*

To determine the effect of amino acid substitutions in HP and also to assess whether or not the normal C terminus of HP is important, several site-specific mutations in *hyp* were made according to the Kunkel procedure using a commercially available mutagenesis kit (BioRad®). Primers (18 to 24 mers) were synthesized (Texas A&M Biology Department) with the appropriate base modifications to result in the following conversions: amino acid 21: asparagine to aspartic acid; amino acid 25: proline to alanine; amino acid 32: valine to aspartic acid; and amino acids 42 and 43: aspartic acid, aspartic acid to tyrosine, tyrosine. The mutations induced were verified by dideoxy sequencing and the respective *hyp* DNA sequences subcloned in frame at the position of the C terminus of β-galactosidase in the vector pUR278.

The amber stop codon of *hyp* (TAG) was mutated to TAC (tyrosine) in pUC19:*hyp*, where *hyp* is inserted between the *Bam*HI and *Sal*I sites of pUC19. After excision with *Bam*H1/*Sal*I and 7.5% PAGE, the 145-bp band was excised and cloned into the *Bam*H1/*Sal*I sites of pUR278. This construct codes for a 1085-amino acid fusion polypeptide containing the complete sequence of LacZ (1024 amino acids) and HP (43 amino acids) plus a C terminal extension of 15 amino acids whose sequence is anticipated to be YESTLESLSMISCQT.

Construction and Properties of *lacZ:hyp:cam*

pUC8 containing *cam* (pUC8:cam) was obtained from Dr. D. Watterson and transformed into *E. coli* MB392. After cleavage of plasmid miniprep DNA with *Eco*R1 and *Bam*H1 and electrophoresis, *cam* DNA was extracted from 7.5% polyacrylamide gels with a buffer of 0.5 *M* ammonium acetate, 10 m*M* magnesium acetate, 0.1 m*M* EDTA, and 0.1% SDS. The single-stranded ends were filled in with T4 polymerase.

pUR278:ptz19:hyp plasmid DNA was cleaved with *Hind*III and *Sal*I and the single-stranded ends were filled with T4 polymerase. The DNA was dephosphorylated with seven units of calf intestine alkaline phosphorylase at 50°C for 1 h.

Overnight, 2 μg *cam* DNA and 0.2 μg of the linearized vector were ligated with T4 ligase at room temperatures. The DNA was incubated with *Xba*I to cut recircularized vector DNA not containing *cam*. After transformation into *E. coli* MB392 and ampicillin selection, most colonies carried a *cam* insert in the vector. The two possible orientations were distinguished electrophoretic analysis of *Bam*HI/*Sal*I and *Bam*H1/*Hind*III cut DNA, the correct orientation yielding a 325- and a 505-bp fragment, respectively, and the inverted orientation yielding a 420- and a 240-bp fragment, respectively.

Synthesis of the fusion polypeptide was induced, cells were lysed by passage through a French pressure cell (16,000 psi), and the lysate was centrifuged at 45,000 rpm for 1 h at 4°C. The membranous pellet was solubilized in Tris-acetate buffer containing 2% deoxycholate and purified by affinity chromatography using *p*-aminophenyl-1-thio-β-D-galactoside sepharose.

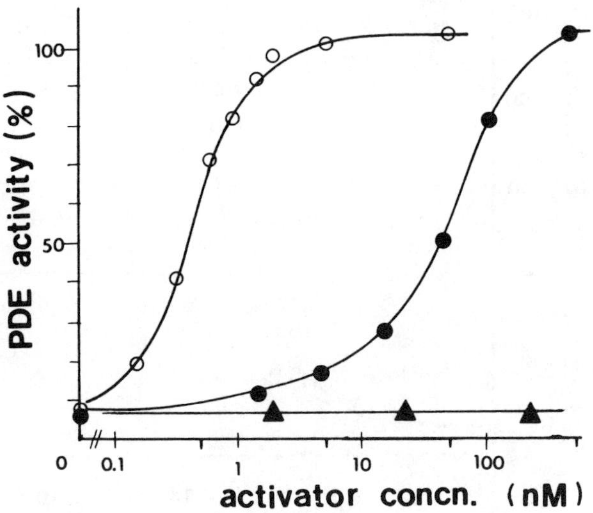

Figure 2. The activation curve of PDE by native Cam (-○-) and hybrid protein: LacZ:HP, (-△-); LacZ:HP:Cam (-●-). Half-maximal PDE activation required about 0.4 n*M* native Cam or 40 n*M* of LacZ:HP:Cam. The hybrid protein LacZ:HP did not activate PDE.

Activity of the Triple Fusion Protein

LacZ:HP:Cam retains both β-galactosidase activity and calmodulin activity. Calmodulin activity was measured by the method of Schiefer[21] using phosphodiesterase (PDE) as a target. Briefly, cyclic adenosine monophosphate (cAMP) is hydrolyzed to AMP by PDE activated by calmodulin. The AMP formed is dephosphorylated to adenosine by alkaline phosphatase and adenosine is converted to inosine by adenosine deaminase. The change in absorbance at 265 nm is proportional to calmodulin activity in the sample. Basal activity of PDE in the absence of calmodulin is only 8% of the maximally stimulated activity.

LacZ:HP:Cam stimulated PDE to the same maximal levels as native calmodulin. However, the activator concentration required for half-maximal activation of PDE was 40 n*M* for LacZ:HP:Cam and only 0.4 n*M* for native calmodulin (Figure 2). LacZ:HP was inactive. Since calmodulin and PDE are believed to associate in the presence of Ca^{+2}, it seems possible that this interaction is sterically hindered by the presence of the bulky β-galactosidase or possibly by the hydrophobic peptide in LacZ:HP:Cam. Activation is Ca^{+2} dependent. The reaction medium contained 0.3 m*M* Mg^{+2}, 0.8 m*M* EGTA, 250 n*M* native calmodulin or LacZ:HP:Cam, and various amounts of total Ca^{+2}. Free Ca^{+2} concentrations were determined with a computer program using a Ca^{+2}: EGTA association constant of $10^{10.97}$.[22] As shown in Figure 3, Ca^{+2} activation is highly cooperative and about the same for both for native calmodulin and LacZ:HP:Cam.

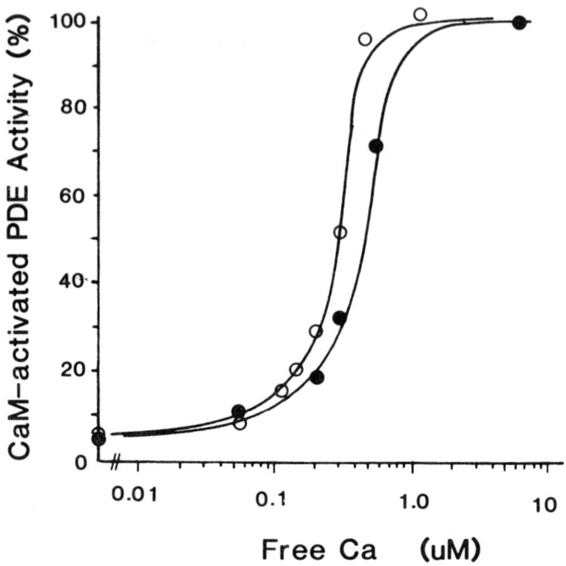

Figure 3. Ca^{2+}-dependent activation of PDE by native Cam (-○-) and hybrid protein LacZ:HP:Cam (-●-). The reaction medium contained 0.3 m*M* Mg, 0.8 m*M* EGTA, and 250 n*M* native Cam or LacZ:HP:Cam.

INTRACELLULAR LOCATION OF THE FUSION POLYPEPTIDES IN *E. COLI*

We anticipated that the fusion protein might be found *in vivo* on the inner aspect of the *E. coli* inner membrane, since the cytochrome b_5 sequence should direct the protein to the membrane. There is no reason to expect that the cytochrome b_5 sequence should result in transport of the protein, especially since β-galactosidase to which has been added a signal sequence cannot be transported through the membrane. However, if the cytochrome b_5 sequence were recognized as abnormal and removed, the β-galactosidase portion of the fusion protein might be found in the cytoplasm.

The initial experiments were conducted with the plasmid construct, pSKG2, which results in an enzymatically inactive polypeptide. Plasmids were transferred into *E. coli* strains MB392 or MC4100 (Table 1). Cells carrying the fusion protein from pSKG2 were grown to an O.D. of 0.3 in 1-l volumes, induced with 1 m*M* IPTG, grown to an O.D. of 1.0, centrifuged, and resuspended in 10 ml Tris-acetate buffer at 4°C. To the washed cells in Tris-acetate buffer, pH 8.1, 100 μg/ml PMSF was added and the cells were immediately lysed twice in the French press. Unlysed cells and debris were removed by centrifugation at 2,000 × *g* for 10 min and the lysate was divided into supernatant and membrane fractions by centrifugation at 180,000 × *g* for 1 h at 4°C in the VTi 70 rotor. Most of the ω fragment in MB392 was

TABLE 1
Bacteria and Plasmids

	Bacterial strains:
MB392	F′ *lac*Iq *ZM15 proAB traD365 hsdR514* ($r_k^- m_k^+$), *supE44, supF58,* (*argF-lac*)U169, *galK2, galT22, metB1, trpR55, pro*::Tn5 kan^R
MC4100	F$^-$, *araD139,* (*argF-lacU169*), *rpsL150, relA1, flbB5301, deoC1, ptsF25, rbsR*
Plasmids:	
pMC81	Ap^R, *trpC,B,AlacZ,Y,A* 25,600 bp
pBH20	Ap^R, Tc^R, *lacPO* 4570 bp
pSKG1	pBH20 carrying the *trpC,B,AlacZ,Y,A* segment from pMC81 and lacking a 346-bp segment containing part of Tc
pSKG2	pSKG1, *hyp* inserted ta the *Eco*RI site in *lac*Z
pUR278	*lacPOZ*, Ap^R
pUR278:*hyp*	pUR278 with *hyp* inserted between the *Bam*HI-*Sal*I sites at the end of *lac*Z

found in the supernatant, although some was found in the membrane pellet. The supernatant from MB392\pSKG1 contained most of the ω fragment and β-galactosidase, although some was found in the membrane pellet.

To recover the protein from the membrane, the pellet was extracted with various detergents. Nearly all the protein reactive with anti-β-galactosidase antibody was solubilized with SDS; other detergents were less effective. Deoxycholate was more effective than CHAPS or Triton® X100 or Triton® X114 and it was found that two extractions with 5-ml volumes of 2% sodium deoxycholate in Tris-acetate buffer with intermittent sonication on ice allowed most of the protein to be recovered. Debris was removed by centrifugation at 65,000 × *g* for 1 h at 4°C and detergent was removed by dialysis.

By Western blot of a 10% SDS-PAGE gel using monoclonal anti-β-galactosidase antibody (ProMega), two proteins, both having a higher molecular weight than β-galactosidase, were found in the membrane pellet from MB392\pSKG2. One was located at about the expected position, 119 kDa, for the fusion protein and the other had a molecular weight of about 117 kDa, slightly greater than normal β-galactosidase (116 kDa). It seemed possible either that the smaller protein might be a degradation product of the larger, or that the larger polypeptide resulted from suppression of the *hyp* termination codon in MB392. The 119-kDa protein was not found in MC4100\pSKG2, where the ω fragment is absent (but which also lacks supE44), but the 117-kDa protein is present. One possibility was that, in MB392\pSKG2, the ω fragment, which contains an amino-proximal deletion, protected from degradation a fraction of the fusion protein, which contains a deletion of β-galactosidase sequences at the carboxyl terminus. β-Galactosidase is a tetramer of four identical subunits, and complementation to form active enzyme is well known to occur between nonoverlapping promoter proximal and promoter distal deletions. The complemented forms of β-galactosidase are stable *in vivo* and are not easily dissociated *in vitro*.[23] The observation that approximately equal amounts of the ω fragment and the 119-kDa protein were present was consistent with formation of a complex. Direct binding of the ω fragment to the fusion protein was also suggested by the observation that most of the ω fragment in MB392\pSKG2 was also found in the membrane pellet fraction, whereas in the absence of pSKG2 it was found in the supernatant. Colonies of MC1400\pSKG2 were white on X-gal, showing that the degradation fragment is inactive, but colonies of MB392\pSKG2 were blue, indicating either that the read-through product was active or that complementation was occurring, either with the 119-kDa protein or the 117-kDa protein. Since both the 119- and 117-kDa proteins were found in membrane pellet fraction, but not the supernatant (behavior which suggested that the C terminal cytochrome b_5 HP was present), it seemed likely that the both species contained HP more or less complete at the C terminus and so must differ either at the amino terminus or by read-through extension at the carboxyl terminus. An ochre mutant of β-galactosidase lacking only the C terminal ten amino acids causes the initial 116-kDa species to be degraded to a 90-kDa species which still has the N terminus and α-complementing activity.[24] Two other termination mutations also resulted in a 90,000-mol wt species.[25] This proteolytic cleavage site seems to be masked in these fusion proteins. Presumably the substitution of a new carboxyl terminus in β-galactosidase restores protection to the proteolytic site normally protected by the last ten amino acids.

Since the construct from pSKG2 is an enzymatically inactive protein, it was clear that the experiments and their interpretation would be more straightforward if an enzymatically active polypeptide could be constructed, so that most of the complications discussed above could be eliminated. All further experiments were conducted using pUR278 as the vector and experiments similar to those described above were done, following β-galactosidase enzymatic activity. Most of the enzymatic activity of the fusion protein, as well as protein detected by anti-β-galactosidase antibody or directly on gels, was associated with the membrane fractions (Table 2).

Immunocytochemical analysis of thin-sectioned bacteria demonstrated that the LacZ:HP fusions were associated with the membrane, whereas native LacZ was distributed uniformly throughout the cytoplasm.[26]

The distribution of β-galactosidase activity was also determined for the three LacZ constructs in which HP is located at the carboxyl terminus (HP_C), internally, 15 amino acids removed from the carboxyl terminus (HP_I), and near the N terminus (HP_N). In all three instances, most of the enzymatic activity was associated with the membrane (Table 3). This suggests strongly that a terminal carboxyl group is not necessary for the membrane-binding function of HP, although it is evident that when HP is located towards the amino terminus a smaller percentage of the protein is membrane bound. The apparent increase in the cytoplasmic fraction for this fusion polypeptide did not result from proteolytic removal of HP, as SDS-PAGE and Western blots demonstrated that HP_N from the membrane and supernatant fractions had equal molecular weights.

TABLE 2
Expression and Distribution of β-Galactosidase and Its Derivative Fusion Proteins

Plasmid	Expressed protein	Molecular weight of the protein (kDa)	β-Galactosidase activity			Percentage in membrane fraction
			Total (ku/l-broth cell)	In cytoplasm	In membrane fraction	
pUR278	LacZ	116	8384	7320	1064	13%
pUR278:HP	LacZ:HP	121	7130	480	6650	93%
pUR278:ptz19:HP	LacZ:HP:C15	123	5160	640	4520	87%
pUR278:HP:Cam	LacZ:HP:Cam	138	9200	2400	6800	74%

TABLE 3
Distribution of β-Galactosidase Activity for Different HP Locations

Polypeptide	Percent in membrane pellet	Percent in supernatant
LacZ (pUR278) (n = 7)	90.6	9.4
LacZ:HP^C (n = 7)	9.4	90.6
LacZ:HP_I (n = 6)	11.1	88.9
LacZ:HP_N (n = 4)	40.4	59.6

TABLE 4
Distribution of β-Galactosidase Activity for Different Site-Specific Mutations

	Cytoplasm (%)	Membrane (%)
LacZ:$HP_{21\ (N \rightarrow D)}$	22.7	77.3 (n = 7)
LacZ:$HP_{25\ (P \rightarrow A)}$	16.0	84.0 (n = 5)
	8.0	92.0 (n = 1)
LacZ:$HP_{32\ (V \rightarrow D)}$	5.0	95.0 (n = 1)
LacZ:$HP_{42,43\ (DD \rightarrow YY)}$	11.8	88.2 (n = 5)
LacZ:$HP_{W.T.}$	9.4	90.6 (n = 7)
	9.0	91.0 (n = 1)

To be sure of the conclusion that HP located at internal sites can bind the protein to membranes, a large polypeptide, calmodulin, was added to the carboxyl side of LacZ:HP. Like LacZ:HP, LacZ:HP:Cam expressed in *E. coli* is found largely in association with the bacterial membrane after cell lysis (Table 2).

To study the structural features of the HP domain that are essential for membrane association, several specific mutations in *hyp* were constructed by site-directed mutagenesis and confirmed by dideoxy sequencing. Analysis of the distribution of these proteins into membrane and cytoplasmic fractions was made as discussed previously. As shown in Table 4, each of the altered polypeptides retained the ability to associate with the membrane.

ASSOCIATION WITH LIPOSOMES AND BACTERIAL MEMBRANES

Both cytochrome b_5 and LacZ:HP spontaneously associate with liposomes.[27] LacZ:HP:Cam also has this property. LacZ:HP, LacZ:HP:C15, LacZ:HP:Cam, and β-galactosidase were incubated with phosphatidylcholine SUV for 1 h at 37°C and then introduced under a 20 to 50% sucrose gradient. After centrifugation for 4 h at 24,000 × *g* the gradient was fractionated into 0.5-ml aliquots and assayed for β-galactosidase activity and for phospholipid content. A majority of the activity for each of the fusion proteins containing

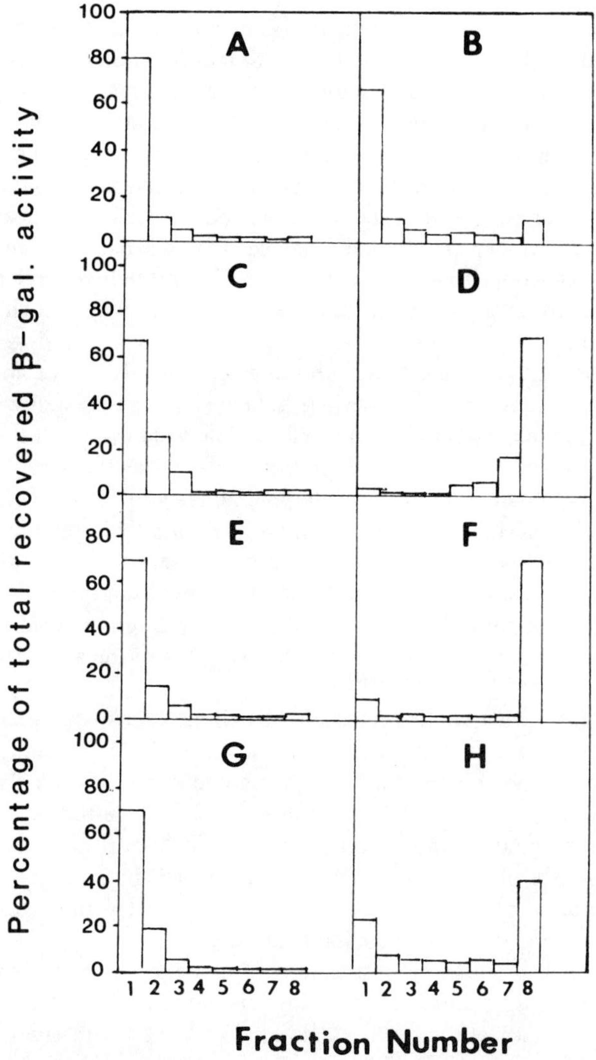

Figure 4. Liposome association of lacZ (A, B), LacZ:HP (C, D), LacZ:HP:C15 (E, F) and LacZ:HP:Cam (G, H). The proteins were incubated in the absence of liposomes (A, C, E, G) or the presence of 200 μl egg-PC liposomes (B, D, F, H) for 1 h at 37°C. The protein was centrifuged at 240,000 × *g* for 4 h in 20 to 50% sucrose density gradients. Fraction 8 in B, D, F, H represents the liposome binding protein in gradient while the free protein remains at the bottom (fractions 1 and 2).

HP was found at the top of the flotation gradient in association with the liposomes (Figure 4). Flotation of LacZ:HP:Cam was less efficient than for the other HP-containing proteins.

Direct interaction of cytochrome b_5 with red cells was initially demonstrated using protein radioactively labeled with the Bolton and Hunter reagent.

Cytochrome b_5 labeled directly on tyrosines with radioactive iodine did not associate with red cells. This may be due to modification of tyrosine residues in the hydrophobic region and suggests that the interaction is dependent on the hydrophobic region. These experiments were done with an early preparation of cytochrome b_5 which contained other contaminating proteins which also labeled well with the Bolton and Hunter reagent. None of these proteins except cytochrome b_5 associated with the red cells. This suggests that the interaction of cytochrome b_5 with red cells was relatively specific, since proteins which copurified with cytochrome b_5 through several purification steps and which must have some features in common with cytochrome b_5 did not associate.

We have demonstrated that both the enzymatically active and the enzymatically inactive forms of chimeric β-galactosidase associated with red cells to a much greater extent than normal β-galactosidase. In the case of the enzymatically inactive species, association was demonstrated by Western blotting of the membrane proteins after hypotonic lysis of the red cells. Both the fusion protein and the ω fragment which lacks the hydrophobic peptide associated with red cells, presumably as a consequence of the formation of a tetrameric structure containing both the ω fragment and the fusion species. Neither β-galactosidase lacking the HP nor the ω fragment in the absence of the fusion protein were associated with red cells as measured by Western blots.

Association was demonstrated also by direct enzymatic assay using the enzymatically active fusion protein. As shown in Table 5, chimeric β-galactosidase having the cytochrome b_5 hydrophobic domain is mostly associated with the red cells after a 24-h incubation in phosphate-buffered saline (pH 6.0) containing glucose and inosine. Only 5 to 10% of β-galactosidase coded for by the parent plasmid pUR278 and fractionated in a similar manner into either a supernatant fraction or a membrane pellet fraction was associated with the red cells after a similar incubation.

We have also followed association of cytochrome b_5 or other proteins and red cells using fluorescent proteins labeled with Texas Red. These observations also indicate that the association observed with fluorescent cytochrome b_5 is the result of a specific interaction. For example, red cells become very fluorescent after incubation with cytochrome b_5 at concentrations of 25 μg/ml or less, but do not label with other proteins at the same concentrations. Red cells can be made faintly fluorescent after incubation with high concentrations (0.2 to 2 mg/ml) of fluorescent proteins, for example, commercial β-galactosidase at concentrations greater than 1 mg/ml. However, binding of the other proteins, but not cytochrome b_5, can be greatly reduced or prevented by addition of HSA during incubation, and the other proteins, but not cytochrome b_5, can be partly or completely removed by high salt washes (1 to 2.5 *M* NaCl). These other proteins, particularly after modification by addition of very hydrophobic residues of Texas Red, perhaps bind to the red cells by a combination of ionic interactions between positively charged

TABLE 5
Association of β-Galactosidase with Red Cells

Protein	Source	pH	Percent activity associated with red cells
Fusion	Membranes	6.0	62
		8.0	7
	Supernatant	6.0	24
		8.0	17
Normal	Membranes	6.0	9
		8.0	0.3
	Supernatant	6.0	5
		8.0	2

Note: Fusion protein contains the entire amino acid sequence of β-galactosidase and continues with the HP sequence of cytochrome b_5 attached at the carboxyl terminus, prepared by cloning *hyp* behind *lac*Z in pUR278. The gene was expressed in MB392 and the protein separated by centrifugation into membrane and supernatant fractions after disruption in a French press. Normal β-galactosidase is that protein synthesized from the parental plasmid. Little or no binding was observed for commercial β-galactosidase.

amino acids and sialic acid residues and hydrophobic interactions between Texas Red residues and the membrane.

We have fluorescently labeled rabbit red cells with either cytochrome b_5 or high concentrations of rabbit hemoglobin and have reinfused these into rabbits. We were surprised to find that the red cells labeled with fluorescent hemoglobin circulated for several days and the fluorescent label remained associated with the membrane after hemolysis and resealing. This observation suggests that proteins, perhaps modified by addition of hydrophobic groups, might bind well enough to red cells to accomplish some of the same objectives envisioned for the HP fusion proteins. Preliminary experiments with cytochrome b_5-labeled red cells indicate that these fluorescent cells survive better after reinfusion.

Using cytochrome b_5 fluorescently labeled with FITC, membrane mobility studies were conducted in collaboration with Dr. Reiner Peters (Max-Planck-Institut, Frankfurt, Germany) using the fluorescence reactivation after a photobleaching technique. After photobleaching a portion of the cellular membrane, the rate at which fluorescent proteins move into the bleached area was determined. The mobility of the protein in the membrane can be calculated and the results for cytochrome b_5 indicate that the mobile fraction is about 50% and has a lateral diffusion coefficient of about 1 to 8 $\times$ 10^{-9} cm^2/s in erythrocytes and 3T3 cells.[28] This observation constitutes evidence that the fluorescent cytochrome b_5 protein is mobile on or in the membrane.

Based on enzyme activity, we estimate about 100,000 molecules of the β-galactosidase fusion protein are associated per red cell, and based on fluorescence we estimate about 150,000 molecules of cytochrome b_5 are associated. When red cells with enzymatically active β-galactosidase are plated on agar containing X-gal, the result is blue red cells.

ACKNOWLEDGMENTS

The expert technical assistance of Lydia Najera, Ramona Sandoval, Lee Ann Benson, and Zhi-Yun Lu is gratefully acknowledged. This research was supported by a grant from the Texas Advanced Technology Research Program.

REFERENCES

1. Ihler, G. M., Glew, R. W., and Schnure, F. W., Enzyme loading of erythrocytes, *Proc. Natl. Acad. Sci. U.S.A.*, 70, 2663, 1973.
2. Beutler, E., Dale, G. L., Guinto, E., and Kuhl, W., Enzyme replacement therapy in Gaucher's therapy in Gaucher's disease: preliminary clinical trial of a new enzyme preparation, *Proc. Natl. Acad. Sci. U.S.A.*, 74, 4620, 1977.
3. Beutler, E., Gaucher's disease, *N. Engl. J. Med.*, 325, 1354, 1991.
4. Berman, J. D., Gallalee, J. V., Williams, J. S., and Hockmeyer, W. D., Activity of pentamidine-containing human red cell ghosts against *Visceral leishmaniasis* in the hamster, *Am. J. Trop. Med. Hyg.*, 35, 297, 1986.
5. DeLoach, J., Peters, S. P., Pinkard, O., Glew, R., and Ihler, G. M., Effect of glutaraldehyde treatment on enzyme-loaded erythrocytes, *Biochim. Biophys. Acta*, 496, 507, 1977.
6. Sprandel, U., Hubbard, A. R., and Chalmers, R. A., Survival of carrier erythrocytes in dogs, *Clin. Sci.*, 59, 7P, 1980.
7. Sprandel, U., Hubbard, A. R., and Chalmers, R. A., Towards enzyme therapy using carrier erythrocytes, *J. Inher. Metab. Dis.*, 4, 99, 1981.
8. Ihler, G. M., Erythrocyte carriers, in *Methods of Drug Delivery, International Encyclopedia of Pharmacology and Therapeutics*, Ihler, G. M., Ed., Pergamon Press, Oxford, 1986, chap. 1.
9. Lewis, D. A. and Alpar, H. O., Therapeutic possibilities of drugs encapsulated in erythrocytes. *Int. J. Pharm.*, 22, 137, 1984.
10. Eichler, H. G., Rameis, H., Bauer, K., Korn, A., Bacher, S., and Gasic, S., Survival of gentamicin-loaded carrier erythrocytes in healthy human volunteers, *Eur. J. Clin. Invest.*, 16, 39, 1986.
11. Ihler, G. M., Lantzy, A., Purpura, J., and Glew, R., Enzymatic degradation of uric acid by uricase-loaded human red blood cells, *J. Clin. Invest.*, 56, 595, 1975.
12. Updike, S. J., Wakamiya, R. T., and Lightfoot, E. N., Asparaginase entrapped in red blood cells: action and survival, *Science*, 193, 681, 1976.
13. Ataullakhanov, F. I., Vitvitskii, V. M., Zhabotinskii, A. M., and Pichugin, A. V., Permeability of human erythrocytes to asparagine, *Biokhimiya*, 50, 1733, 1985.

14. Verner, K. and Schatz, G., Protein translocation across membranes, *Science*, 241, 1307, 1988.
15. Roessner, C., Struck, D., and Ihler, G. M., Injection of DNA into liposomes by bacteriophage lambda, *J. Biol. Chem.*, 258, 643, 1983.
16. Roessner, C. A., Struck, D. K., and Ihler, G. M., Transfer of DNA across membranes: a model system using liposomes, *Biotechniques*, 140, 1984.
17. Greenhut, S. F., Bourgeois, V. R., and Roseman, M. A., Distribution of cytochrome b_5 between small and large unilamellar phospholipid vesicles, *J. Biol. Chem.*, 261, 3670, 1986.
18. von Bodman, S., Schuler, M. A., Jollie, D. R., and Sligar, S. G., Synthesis, bacterial expression, and mutagenesis of the gene coding for mammalian cytochrome b_5, *Proc. Natl. Acad. Sci. U.S.A.*, 83, 9443, 1986.
19. Grantham, R., Gautier, C., and Gouy, M., Codon frequencies in 119 individual genes confirm consistent choices of degenerate bases according to genome type, *Nucleic Acids Res.*, 8, 1893, 1980.
20. Ruther, U. and Muller-Hill, B., Easy identification of cDNA clones, *EMBO J.*, 2, 1791, 1983.
21. Schiefer, S., Calmodulin, in *Methods of Enzymatic Analysis*, Vol. 9, 3rd ed., Bergmeyer, H. U., Ed., VCH Verlagsgesellschaft, Weinheim, 1985, 317.
22. Goldstein, D., Calculation of the concentration of free cations and cation-ligand complex in solutions containing multiple divalent cations and ligands, *Biophys. J.*, 26, 235, 1979.
23. Villarejo, M., Zamenhof, P. J., and Zabin, I., β-Galactosidase. *In vivo* α-complementation, *J. Biol. Chem.*, 247, 2212, 1972.
24. Kowit, J. D. and Goldberg, A. L., Intermediate steps in the degradation of a specific abnormal protein in *Escherichia coli*, *J. Biol. Chem.*, 252, 8350, 1977.
25. McKnight, J. L. and Fried, V. A., Limited proteolysis. Early steps in the processing of large premature termination fragments of β-galactosidase in *Escherichia coli*, *J. Biol. Chem.*, 256, 9652, 1981.
26. George, S. K., Najera, L., Sandoval, R. P., Countryman, C., Davis, R. W., and Ihler, G. M., Hydrophobic domain of cytochrome b_5 is capable of anchoring β-galactosidase in *E. coli* membranes. *J. Bacteriol.*, 171, 4569, 1989.
27. George, S. K., Najera, L., Sandoval, R. P., and Ihler, G. M., β-Galactosidase fused to the hydrophobic domain of cytochrome b_5 spontaneously associates with liposomes, *Biochim. Biophys. Acta*, 1061, 26, 1991.
28. George, S. K., Xu, Y.-H., Benson, L. A., Pratsch, L., Peters, R., and Ihler, G. M., Cytochrome b_5 and a recombinant protein containing the cytochrome b_5 hydrophobic domain spontaneously associate with the plasma membranes of cells, *Biochim. Biophys. Acta*, 1066, 131, 1991.

Chapter

6

Progestasert® and Ocusert® Therapeutic System Technologies

Mary Prevo
ALZA Corporation
Palo Alto, CA

INTRODUCTION

In the early 1970s, the relatively new concept of using precisely controlled drug delivery moved into the pharmaceutical development stage.[1,2] Previously, applications of this concept were mainly confined to intensive-care wards, where the use of electromechanical pumps to meter i.v. infusions made possible the safe administration of such drugs as dopamine, dobutamide, and nitroprusside.[3]

In the mid-1970s, ALZA Corporation introduced products that provide rate-controlled delivery of drugs for routine administration to outpatients.[4,5] This type of drug delivery can reduce toxicity by (1) controlling the rate of drug input and avoiding the peaks and valleys in drug concentration associated with conventional bolus administration and (2) targeting the drug to specific organs. The Progestasert® intrauterine progesterone contraceptive system and the Ocusert® pilocarpine ocular therapeutic system are examples of targeted drug delivery, and they are the first two products ever approved with label specifications for rate and duration of drug delivery.

0-8493-8896-1/93/$0.00 + $.50

THE THERAPEUTIC SYSTEM CONCEPT

The Progestasert® and Ocusert® systems employ technologies developed for a class of dosage forms called therapeutic systems.[6-8] Such dosage forms are defined by preprogrammed delivery of a drug at a rate and for a time period established to meet specific medical needs. Several components are required to carry out this therapeutic program: a drug delivery module consisting of a drug reservoir; a rate controller (often a membrane); a portal for drug release; a platform to house the delivery module; and, of course, a drug. The drug is selected on the basis of its efficacy, pharmacokinetic behavior, and physical-chemical properties, permitting effective transfer from delivery module to target site in the body. Unlike the situation with conventional pharmaceutical products, a short half-life is considered a desirable drug property with rate-contolled administration because it provides the opportunity for close control over drug levels in the body through control of drug input rate. It also permits rapid termination of drug presence in the body, should that become advisable.

THE PROGESTASERT® SYSTEM

The Progestasert® system delivers progesterone (the natural hormone) for 1 year from a T-shaped (36 × 32 mm) intrauterine platform (Figure 1). The platform, which also serves as the rate controller, is an ethylene/vinyl acetate copolymer membrane. It contains a reservoir of progesterone in silicone oil. The energy source for drug release is the gradient of excess hormone contained in the reservoir.

Contraceptive Concept

Oral contraceptives — in order to deliver to target organs the small dose needed for contraception — supply synthetic hormones in amounts sufficient to affect major nonreproductive body systems. Decreasing the dosage and reducing or eliminating the estrogenic component, with the aim of minimizing effects on nonreproductive functions, have resulted in the so-called minidose, progestin-only pill. This approach, however, is still associated with systemic hormonal side effects as well as with higher pregnancy rates compared to combination products.[9]

During use of systemically applied synthetic analogs of progesterone, histologic changes occur in the endometrium that are believed to have an antifertility effect. Similar changes occurred in small-scale controlled studies in which progesterone itself was applied topically to the uterus from silastic capsules.[10-12] This suggested the delivery of progesterone from an intrauterine platform to provide localized hormonal contraception without preventing ovu-

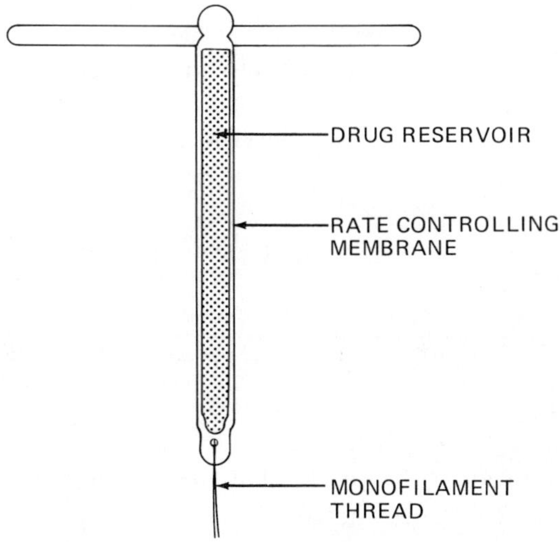

Figure 1. Diagram of Progestasert® intrauterine progesterone contraceptive system (36 × 32 mm).

TABLE 1
Pregnancies among Women Using Placebo Units vs. Progestasert® Systems in Five Clinics*

Termination	Types of units used	Women	Woman months	Number of pregnancies
1-Year study (four centers)	Progestasert® system	370	3048	3
	Placebo unit	330	2953	30
At early removal of placebo	Progestasert® system	98	844	2
(one center)	Placebo unit	99	652	10
Total (five centers)	Progestasert® system	468	3892	5
	Placebo unit	429	3605	40

* First insertions only.

lation or relying on the induction of uterine inflammation. Markedly higher pregnancy rates with placebo Progestasert® systems than with systems delivering progesterone confirmed that the hormone is the source of system efficacy, leading to early termination of placebo use in one center (Table 1). The system is relatively effective; it had a Pearl index of 2.9 in large multicenter clinical trials.[13]

Continuous delivery of progesterone to the target organ permits use of the short-lived natural hormone — which has a half-life of only 45 to 90 min — to provide a year of contraception. To permit once-daily systemic dosing, longer-lived synthetics had to be developed for oral use. The mechanism of action for the hormonally active Progestasert® system includes the local for-

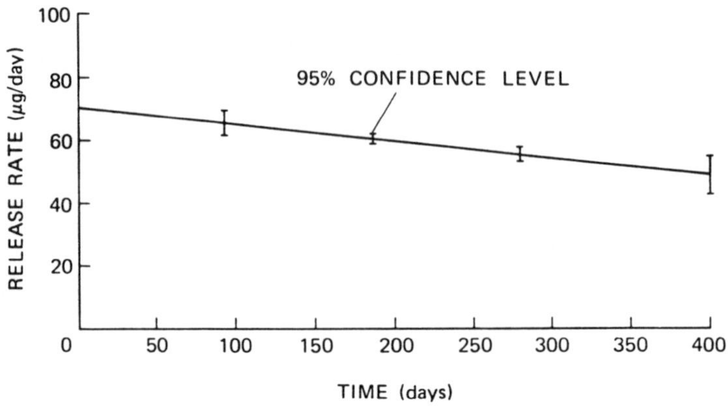

Figure 2. *In utero* progesterone release rate based on 429 systems used in humans for intervals up to 400 d.

eign-body reaction in the endometrium and the progesterone action on the endometrium to produce an environment not conducive to nidation. Compared to inert or copper IUDs, Progestasert® decreases bleeding, cycle irregularity, and dysmenorrhea.[14] The need for annual replacement of Progestasert® ensures that patients will return for a Pap smear, and general gynecological examination can be considered as a benefit of using the system.[15]

System Functionality and Dosage

The *in utero* progesterone release rate of 65 μg/d is predictable for 1 year, being controlled not by patient-dependent biological factors, but by the system itself. The results of ultraviolet (UV) analysis of residual progesterone in 429 systems after varying times *in utero* indicate that the average daily *in utero* release is within ±10 μg of the 65 μg/d value (Figure 2). The system delivers 24 mg progesterone over 1 year. Thus, it achieves endometrial changes sufficient for effective contraception with the amount of progesterone the mature corpus luteum secretes in a single day.[16,17] If the 65 μg of progesterone released each day by a system were to enter the general circulation directly, the hormonal plasma concentration would rise by no more than 0.5 to 5% above the level resulting from endogenous production. This increase is within the range of day-to-day physiologic variations and is not high enough to act systemically on the hypothalamus or pituitary.[18] Endogenous production of the hormone reaches 15 to 50 mg daily in the midluteal stage.

Systemic Safety

The small dosage of progesterone delivered to the endometrium suggested that the Progestasert® system would be unlikely to cause toxic effects or

TABLE 2
Metabolism of ^{3}H-Progesterone by Human Endometrium *In Vitro* During Luteal Phase of Cycle or Treatment with Progestins

Amount unlabeled progesterone added (ng)[a]	Percent ^{3}H-progesterone metabolized per 100 mg tissue[b]
0	10.9, 10.2, 0.8, 12.9, 9.1
20	6.4
200	4.5
2000	5.6, 3.6

[a] ng = nanogram (10^{-9} g).
[b] Each value represents tissue from a different patient; incubation time was 180 min.

TABLE 3
Metabolism of ^{3}H-Progesterone by Human Endometrium *In Vitro* During Follicular Phase of Cycle or Treatment with Estrogens

Amount unlabeled progesterone added (ng)[a]	Percent ^{3}H-progesterone metabolized per 100 mg tissue[b]
0	5.4
20	7.3
200	5.6
2000	5.9

[a] ng = nanogram (10^{-9} g).
[b] Each value represents tissue from a different patient; incubation time was 180 min.

extrauterine hormonal effects. To test whether or not hormonal effects were limited to the uterus, studies were carried out (1) to measure the ability of the endometrium to metabolize exogenous progesterone delivered to the uterus and (2) to detect any extrauterine effects of the exogenous progesterone.

Tissue obtained from six patients scheduled for hysterectomy or dilatation and curretage was prepared and examined for indications other than suspected carcinoma and was evaluated for endometrial metabolism of progesterone. Samples were minced and various quantities of unlabeled progesterone, together with ^{3}H-progesterone, were added. Incubations were done at 37°C. Progesterone was separated from its metabolites by standard paper chromatographic techniques. The experiment showed that the endometrium in both the follicular and the luteal stages of the cycle has the ability to metabolize amounts of progesterone in excess of the 65 μg delivered by the Progestasert® system (Tables 2 and 3).

Systems delivering ^{3}H-progesterone, 80 μg/d, were placed in the uteri of baboons to measure entry of the hormone and its metabolites into systemic blood. Because a mature female baboon weighs about one third as much as

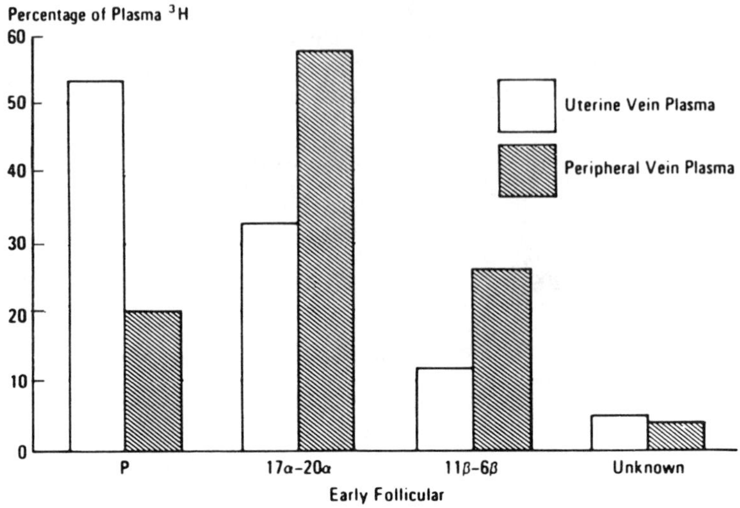

Figure 3. Distribution of progesterone and its metabolites in uterine vein and peripheral vein plasma during follicular stage of the menstrual cycle in a baboon (P = progesterone, 17α-20α = 17α OH and 20α OH progesterone, 11β-6β = 11β OH and 6β OH progesterone).

a mature woman, the potential rate of progesterone delivery into the systemic circulation was, on a relative basis, over three times the rate for a human. Samples of urine, peripheral plasma, and uterine vein plasma, taken at four stages of the menstrual cycle, were purified and assayed for radioactivity. The high level of metabolites in the peripheral uterine circulation supports the hypothesis that significant endometrial metabolism of progesterone occurs when this hormone is delivered into the uterine cavity (Figure 3).

Ermini and co-workers[19] report that histological examination showed decidual changes in endometrium adjacent to the devices that are attributed to the progesterone released from the Progestasert® system. Autoradiography, to elucidate tissue distribution, showed that progesterone is present by the adjacent epithelium, the glands adjacent to the surface, and was well distributed in the stroma and capillary walls, but does not readily penetrate the deeper portion of the endometrium. This supports the hypothesis that progesterone, delivered *in utero* on a continuous basis, exerts primarily a local action.[20-22]

In preclinical and clinical studies, researchers attempted to detect extrauterine effects on the hypothalamic-pituitary-ovarian axis of progesterone delivered to the uterus. This axis is more sensitive to progesterone than any other system in the body. In humans, blood chemistries were also studied to detect extrauterine effects.

In six mature female baboons, following control cycle observations, uterine systems releasing 65 μg/d progesterone were placed. Radioimmune assays

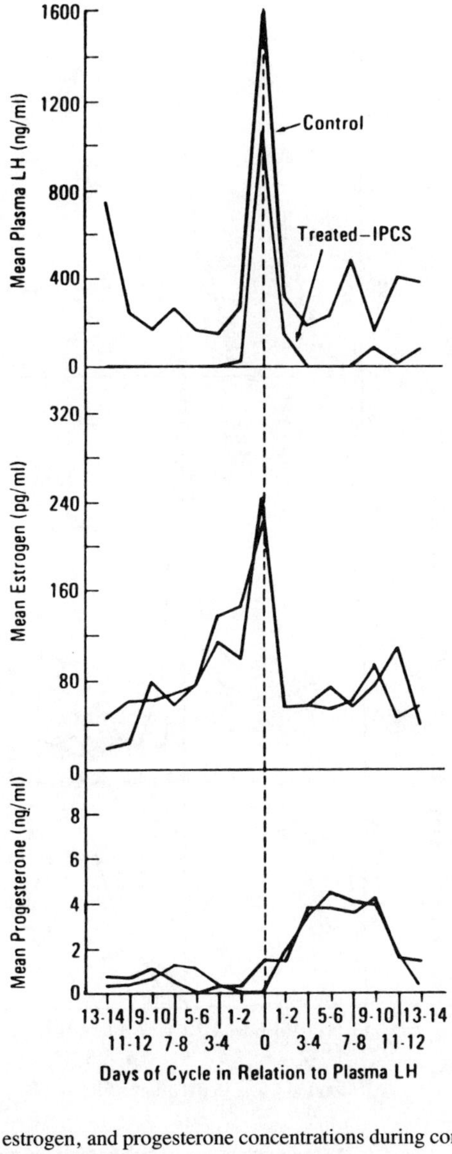

Figure 4. Mean LH, estrogen, and progesterone concentrations during control and progesterone-treated cycles in a representative baboon.

sensitive to submicrogram levels were used to determine cyclic patterns of luteinizing hormone (LH), estradiol, and progesterone in the circulating blood. Average plasma levels of these hormones during control and progesterone treatment cycles (Figure 4) — or during control cycles and cycles with a Lippes Loop in place (Figure 5) — indicated a combined secretion pattern consistent with an ovulatory cycle. Perineal changes also indicated that ovu-

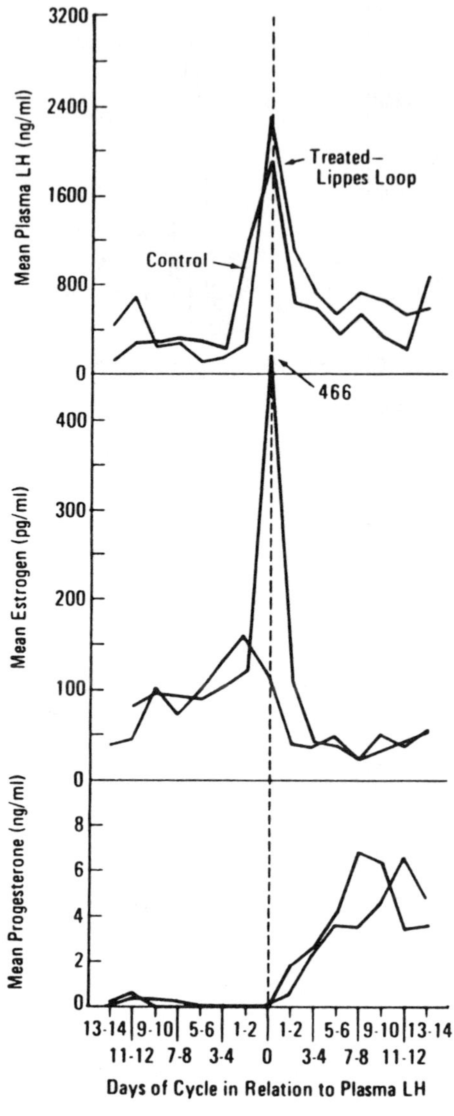

Figure 5. Mean LH, estrogen, and progesterone concentrations during control and Lippes Loop-treated cycles in a representative baboon.

lation continued. In addition, menstrual flow data, as well as duration and frequency of menses, were comparable during control and treated cycles. It was concluded that amounts of progesterone proportionately exceeding amounts that the Progestasert® system delivers in humans exerted no detectable effect on the ovulatory cycle of baboons and were devoid of extrauterine progestational effects.

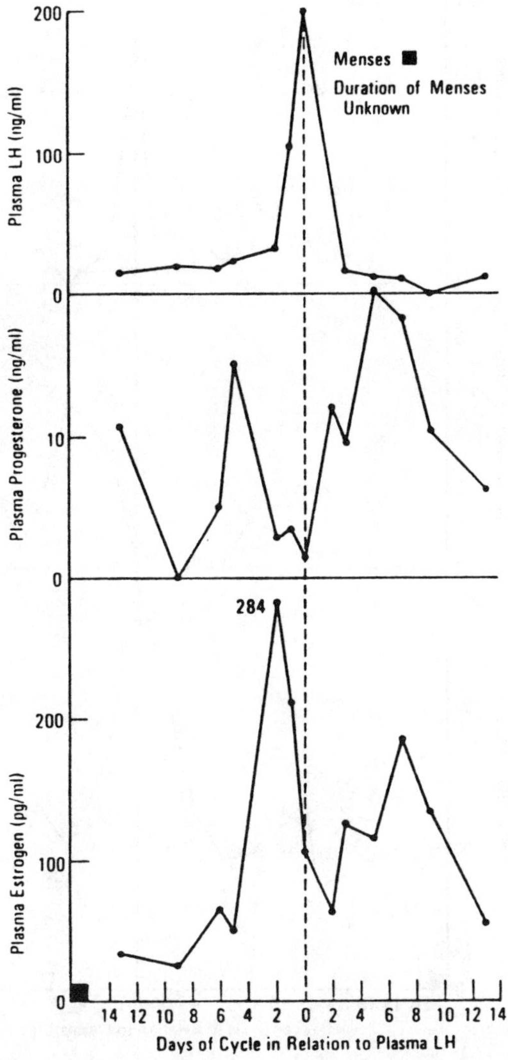

Figure 6. Plasma estrogen, progesterone, and LH concentration in a user of a Progestasert® system.

In eight multiparous women using the Progestasert® system and in four multiparous women using placebo systems for 1 month, pituitary gonadotrophins and steroid hormone levels were assessed during the menstrual cycle. Estrogen, progesterone, and LH patterns were analyzed when blood was drawn at frequent intervals over a 30-d period. Each woman's hormonal responses indicated normal physiologic patterns (Figures 6 and 7). Regulatory hormonal signals and hormonal responses indicated a normal ovulatory cycle.

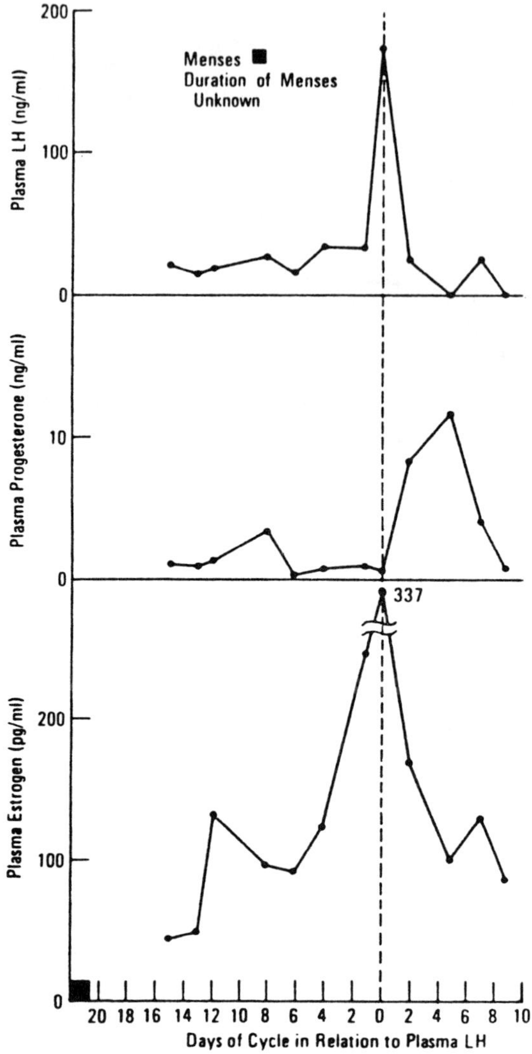

Figure 7. Plasma estrogen, progesterone, and LH concentration in a woman using a unit containing no hormone.

Menstrual data from patient groups in the clinical trials also showed no irregularities that would indicate interference with ovulation (Table 4). The incidence of secondary amenorrhea (defined as cessation of menstruation for twice as long as the patient's usual cycle length) was no greater than that recorded at admission (Table 5).

It was not anticipated that the Progestasert® system would cause changes in physiologic parameters less sensitive to the hormone than the hypothalamic-

TABLE 4
Menstrual Cycles Before and During System Use (Percentage of Women Reporting Various Cycle Lengths)

	<26 d	26 to 29 d	30 to 35 d	>35 d
Admission	6.8	42.8	21.3	29.2
6 Months	9.8	49.1	31.7	9.4
12 Months	10.2	53.2	29.3	7.2

pituitary-ovarian axis. Nevertheless, to detect possible systemic effects, blood chemistry studies were done at intervals during the clinical trials in 150 women; measurements of blood cell counts, hemoglobin, and hematocrit were done in 300. Analysis over time indicated no changes in laboratory values from initial values that were of substantial magnitude or clinical significance (Tables 6 and 7).

THE OCUSERT® SYSTEM

The pilocarpine ocular therapeutic system is a clear, elliptically shaped wafer with a white annular ring measuring approximately 13 × 5.5 × 0.3 mm. When placed in the conjunctival cul-de-sac, it delivers 20 or 40 μg/h of pilocarpine to the eye at an essentially constant rate for 1 week to control intraocular pressure.[23,24] Figure 8 shows the schematic construction of the system. The reservoir of pilocarpine provides a 7-d supply of drug and is enclosed by a thin membrane of ethylene/vinyl acetate copolymer. A retaining ring of the same material, impregnated with titanium dioxide (TiO_2) for visibility, encloses the drug reservoir.

System Functionality and Dosage

Both *in vitro* and *in vivo* evaluation of the pilocarpine delivery rate from the ocular therapeutic systems justify the claim that these products are indeed rate-controlled dosage forms. Figure 9 shows representative *in vitro* results for the 20- and 40-μg/h systems; 24-h intervals were used except in the initial 24-h period. During that time, the release rate went through an initial start-up; thus 6- and 18-h intervals were employed.[25] The *in vivo* release rate can be calculated for single units per time point by removing them from the conjunctival cul-de-sac at differing intervals over their functional lifetimes, and measuring quantity of residual drug. Figure 10 shows the results, demonstrating correspondence between the *in vivo* and *in vitro* situations, which provide the basis for label specifications of therapeutic strength by both drug delivery rate and duration.

TABLE 5
Amenorrhea in Various Patient Groups Before and After System Insertion

Visit	No. women reporting	Amenorrheic lactating women number/%	Amenorrheic former oral contraceptive users[a] number/%	Other amenorrheic number/%
Admission	2918	61/2.0	26/0.9	36/1.2
1 Month	2290	42/1.8	4/0.2	18/0.8
3 Months	1236	17/1.4	7/0.5	12/1.0
6 Months	906	2/0.2	4/0.4	3/0.3
12 Months	383	0/0	11/2.8	3/0.7

[a] Within the 3 months preceding admission to clinical trials.

TABLE 6
Summary of Average Values for Laboratory Blood Chemistry Data

Assay (units)	Admission	1 Month	3 Months	6 Months	12 Months
Glucose (mg %)	89.9	91.1	86.2	87.9	87.8
Uric acid (mg %)	4.3	4.1	4.0	4.1	3.9
Cholesterol (mg %)	206.9	200.9	192.0	188.9	194.8
Thyroxin (μg %)	8.5	8.1	7.6	7.7	8.5
Creatinine (mg %)	0.9	0.9	0.9	0.8	0.8
BUN (mg %)	13.6	13.7	13.3	13.5	14.0
Alk Phos (B-L)[a]	1.4	1.4	1.3	1.3	1.4
Bilirubin (mg %)	0.7	0.6	0.6	0.6	0.6
LDH (B-B)[b]	457.6	449.5	430.9	422.7	430.5
SGOT (S-F)[c]	26.7	28.2	25.4	25.4	24.8
Protein (g %)	7.1	7.0	7.0	6.8	7.1
Albumin (g %)	4.0	4.1	4.0	4.0	4.2
Globulin (g %)	3.0	2.9	2.9	2.8	2.9
A/G ratio[d]	1.4	1.4	1.4	1.4	1.4
Calcium (mg %)	10.0	10.0	10.0	9.9	9.8
Phosphorus (mg %)	3.6	3.8	3.7	3.7	3.5
Chloride (mEq/l)	99.8	99.4	98.5	99.6	98.6
Potassium (mEq/l)	4.2	4.2	4.3	4.0	4.3
Sodium (mEq/l)	139.5	139.1	137.9	138.9	139.1
Sample size	174	144	125	95	53

[a] Alkaline phosphatase (Bessy-Lowry units).
[b] Lactic dehydrogenase (Berger-Broider units).
[c] Serum glutamic oxalate (Sigma-Frankel units).
[d] Albumin/globulin ratio.

TABLE 7
Summary of Average Values for Laboratory Hematology Data

Assay (units)	Admission	1 Month	3 Months	6 Months	12 Months
White blood count ($10^3/mm^3$)	6.9	6.7	7.0	6.8	7.0
Red blood count ($10^3/mm^3$)	4.5	4.5	4.5	4.5	4.4
Hemoglobin (g%)	13.4	13.3	13.4	13.5	13.2
Hematocrit (%)	40.6	40.2	40.6	40.2	39.5
Sample size	2360	296	128	121	264

Intraocular Tissue Levels of Pilocarpine

A comparison of intraocular tissue levels of pilocarpine following administration by eyedrops or by controlled delivery in rabbits is shown in Figure 11. In most tissues, none of which have a direct bearing on hypotensive action, drug levels from the eyedrop regimen were usually severalfold higher than those maintained by continuous delivery with the Ocusert® system. This is particularly true of the lens — a finding of possible significance in view

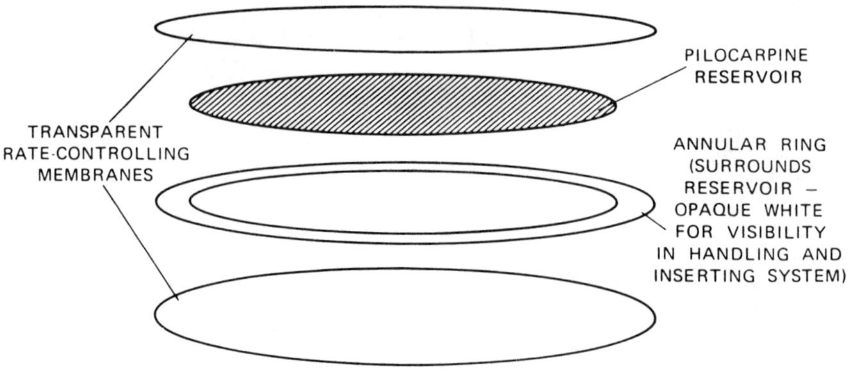

Figure 8. Composition of an Ocusert® ocular therapeutic system that measures approximately 13 × 5.5 × 0.3 mm. (From Place, V. A., Fisher, M., Herbst, S., Gordon, L., and Merrill, R. C., *Am. J. Ophthalmol.*, 80, 706, 1975, Copyright by the Ophthalmic Publishing Company. With permission.)

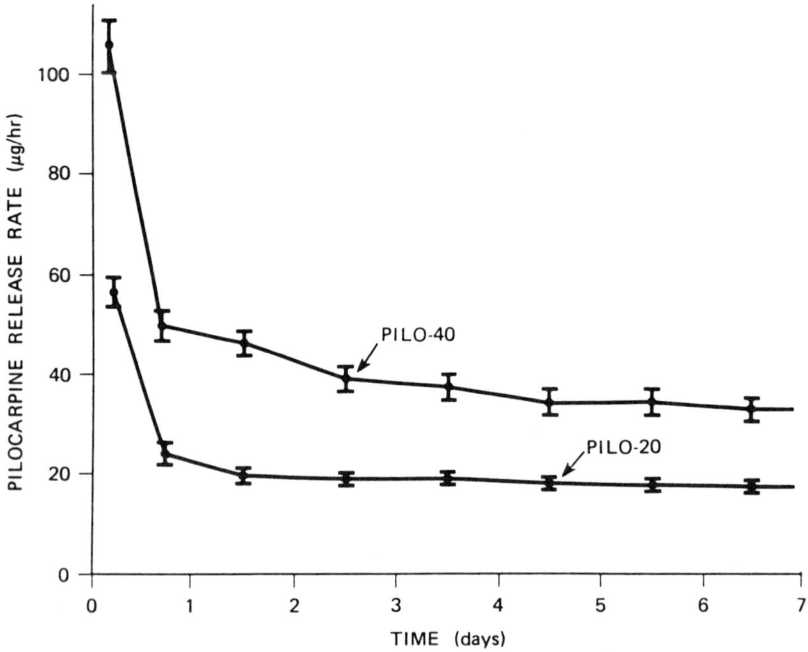

Figure 9. Time course of pilocarpine release *in vitro* from Ocusert® systems, showing near zero-order pattern.

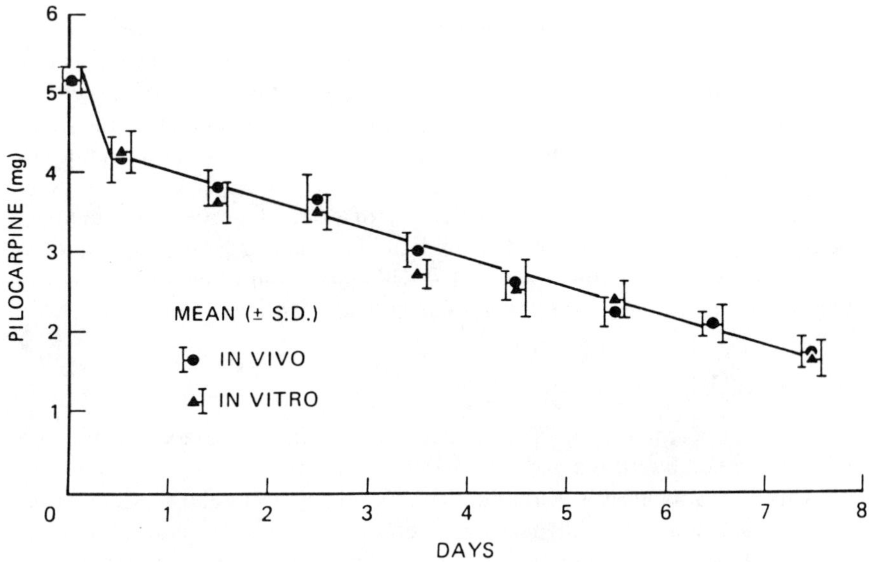

Figure 10. Residual pilocarpine in Ocusert® Pilo-20 systems incubated *in vitro* in saline, or worn by humans, for up to 8 d (each point represents the average of seven Ocusert® systems). Over 7 d *in vitro* units averaged 18.3 ± 1.3 μg/h, and *in vivo* units averaged 17.9 ± 2.9 μg/h.

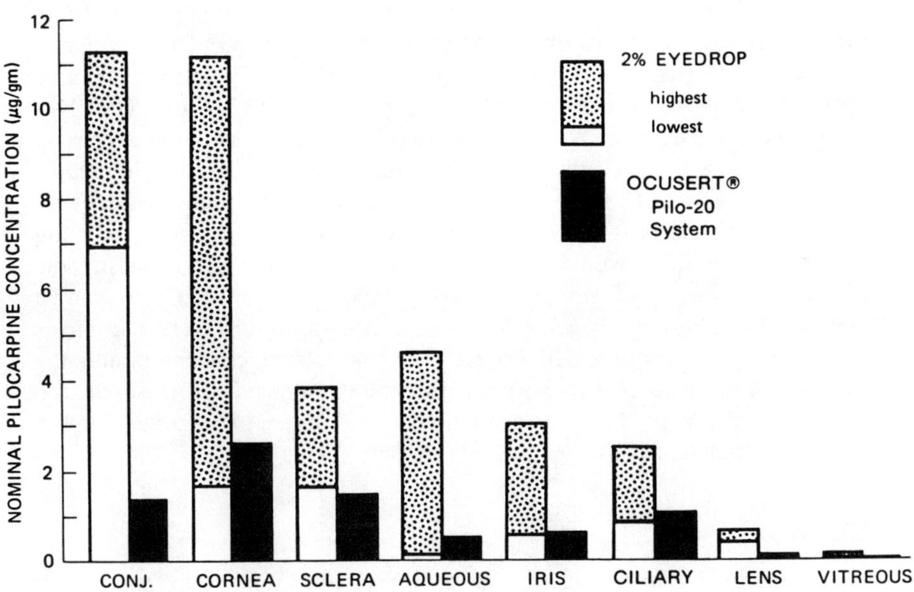

Figure 11. Rabbit ocular tissue levels of ^{14}C derived from ^{14}C pilocarpine.

of the reported cataractogenic activity of pilocarpine — and of the vitreous. In most tissues, the level of eyedrop-delivered pilocarpine never fell below the level maintained by the continuous-delivery systems. A notable exception was the aqueous humor, and presumably it is the relatively rapid turnover of aqueous humor after eyedrop administration that requires eyedrops four times daily.

These findings suggest some element of safety favoring a continuous-delivery system. Pilocarpine delivered in this manner provides long-term residual hypotensive activity associated with a reservoir of the drug in tissues that are not associated with unfavorable side effects from pilocarpine.

Studies of Visual Acuity

With pilocarpine eyedrops, the desired lowering of intraocular pressure is accompanied by miosis and drug-induced accommodation; within a young eye this may amount to several diopters. This cannot be relaxed voluntarily, and therefore may cause myopia for several hours. Miosis produced by pilocarpine can cause visual difficulties, particularly in dim light among elderly patients. In contrast to eyedrops, the Ocusert® system provides to the eye continuous delivery of pilocarpine with equivalent hypotensive efficacy in many patients and induces virtually constant, mild myopia and miosis.[26] Typically, pilocarpine eyedrops are used four times daily. Miosis and myopia are the most severe in the first 30 to 90 min after drop application.

The reduction in miosis and induced myopia in glaucoma patients using the Ocusert® system is described by Brown and co-workers.[26] They compared the following visual effects for eyedrops vs. the Ocusert® system: refractive index, distance vision, near vision, and miosis (Figures 12 to 15). The pharmacological and pharmacokinetic basis for this improvement in the selectivity of pilocarpine action, effected by rate-controlled delivery, is shown in Figures 16 and 17.[25] Figure 16 shows that the slope of the relation between delivery rate and miotic effect is steep in the range between 0 and 60 μg/h, predicting that the miotic effect would be modest at 20 μg/h. In contrast, the myopic effect is independent of delivery rate below 60 μg/h (Figure 17). Also shown here are the relations between pilocarpine concentrations in tears and miosis or myopia. By controlling delivery rate, one controls the concentration in the tears within a range of 2 to 3 μg/ml. In contrast, a standard 50-μl eyedrop of 2% pilocarpine produces concentrations 3 to 4 orders of magnitude higher than those maintained by the Ocusert® system.

Efficacy

The reduction in side effects has to be judged in relation to the strength of the hypotensive action of the 20- and 40-μg/h, rate-controlled administration of the drug. During the first hour after administration of the eyedrops,

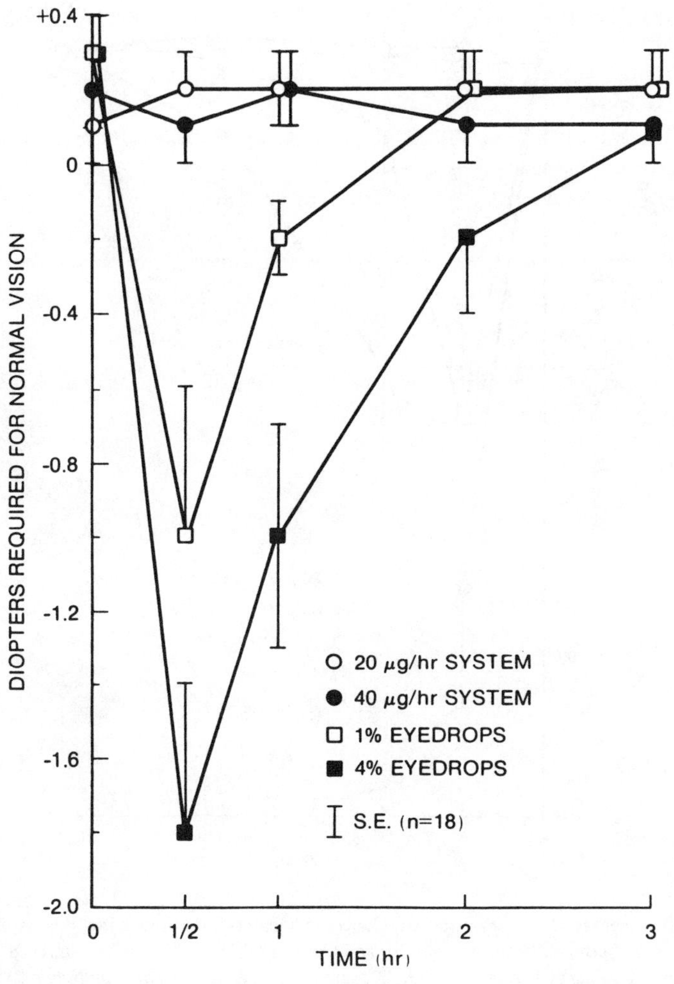

Figure 12. Average refractive errors of glaucoma patients after receiving pilocarpine in 1% eyedrops, 4% eyedrops, Ocusert® Pilo-20, and Ocusert® Pilo-40. (From Brown, H. S., Meltzer, G., Merrill, R. C., Fisher, M., Ferré, C., and Place, V. A., *Arch. Ophthalmol.*, 94, 1716, 1976, American Medical Association. With permission.)

average intraocular pressures decreased by about 2 mmHg from the values obtained 6 h after the last drop, and for the ensuing 2 h they remained fairly constant at 18.1 mmHg (Table 8). Intraocular pressures of patients wearing the 20- or 40-μg/h system remained within a narrow range that averaged 19.2 mmHg throughout the test periods.[26]

Although the amount of drug delivered to the eye from the Ocusert® system was one-fifth that of the eyedrops, the decrease of intraocular pressure was comparable in this group of patients. The systems produced small, con-

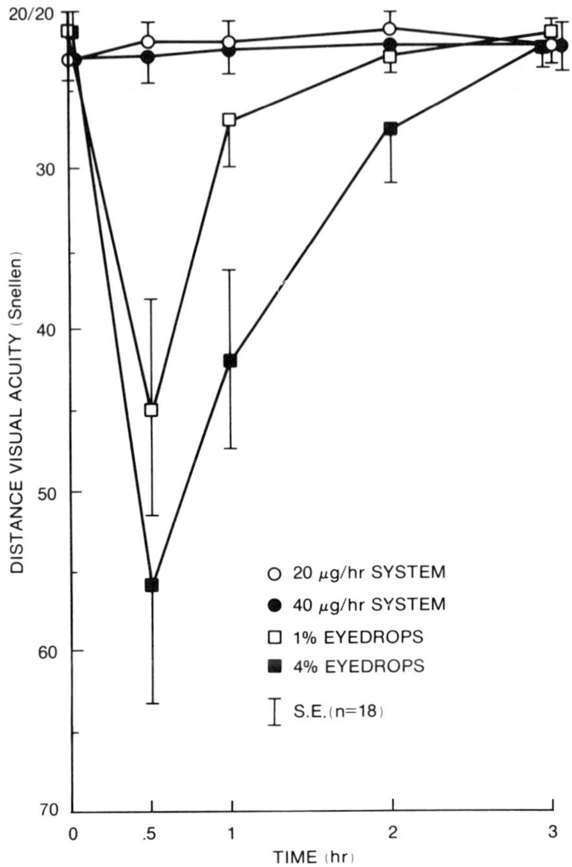

Figure 13. Average distance vision of glaucoma patients after receiving pilocarpine in 1% eyedrops, 4% eyedrops, Ocusert® Pilo-20, and Ocusert® Pilo-40. (From Brown, H. S., Meltzer, G., Merrill, R. C., Fisher, M., Ferré, C., and Place, V. A., *Arch. Ophthalmol.*, 94, 1716, 1976, American Medical Association. With permission.)

stant effects on visual acuity, refractive error, and miosis that did not cause visual handicaps that were not correctable with spectacles.[24,26]

Results from a single group of patients followed for 5 years demonstrated that intraocular pressure remained uniformly reduced during multiyears of pilocarpine treatment with the Ocusert® system and that contraction of ciliary fibers remains virtually constant, as does the drug concentration in ocular tissues.[27] Moreoever, application of placebo units at the end of the study period showed that the reduction in intraocular pressure remained reversible; pressures usually rose within 24 h after discontinuing pilocarpine system use. Hence, continuous administration of pilocarpine to the eye, at a virtually constant rate over 5 years, led neither to clinically recognizable changes in ocular structure nor to any observable reduction in efficacy of the drug.

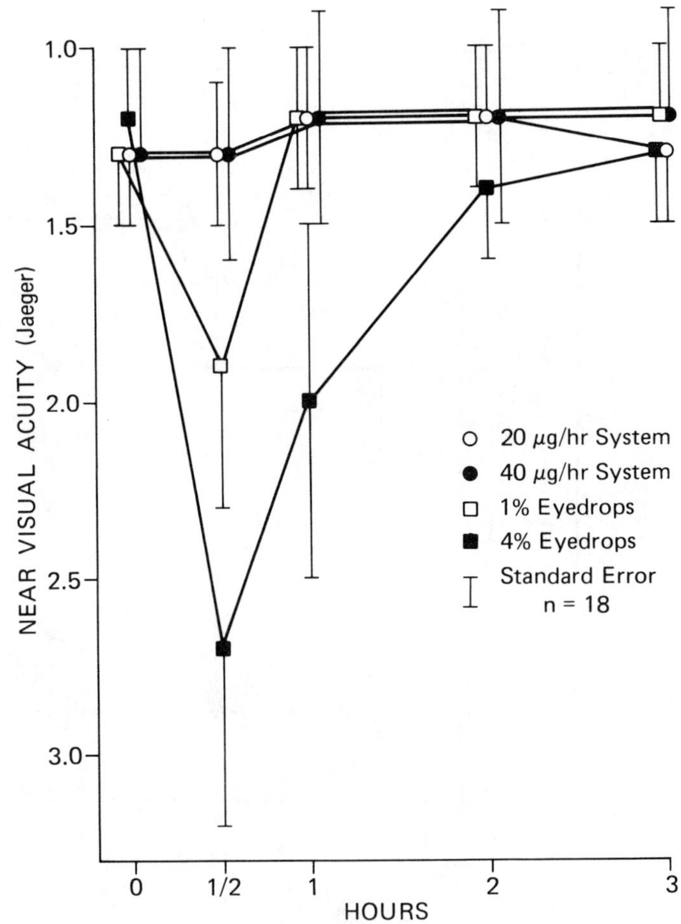

Figure 14. Average visual acuity of glaucoma patients after receiving pilocarpine in 1% eyedrops, 4% eyedrops, Ocusert® Pilo-20, and Ocusert® Pilo-40. (From Brown, H. S., Meltzer, G., Merrill, R. C., Fisher, M., Ferré, C., and Place, V. A., *Arch. Ophthalmol.*, 94, 1716, 1976, American Medical Association. With permission.)

Although the U.S. has shifted away from topical miotics to topical beta-blockers that are effective with only one or two daily instillations, the Ocusert® system has provided a new approach to the treatment of eye disease. This new type of drug delivery avoids the overdosing/underdosing sequence and requires a lower total dose of pilocarpine over 7 d compared to eyedrops[28] (Table 9). The system functions unattended, avoiding loss of protection against the pressure increase that occurs when eyedrops are delayed. Because this system requires less frequent patient intervention for dosing compared to the several high daily doses required in eyedrop administration, it seems likely

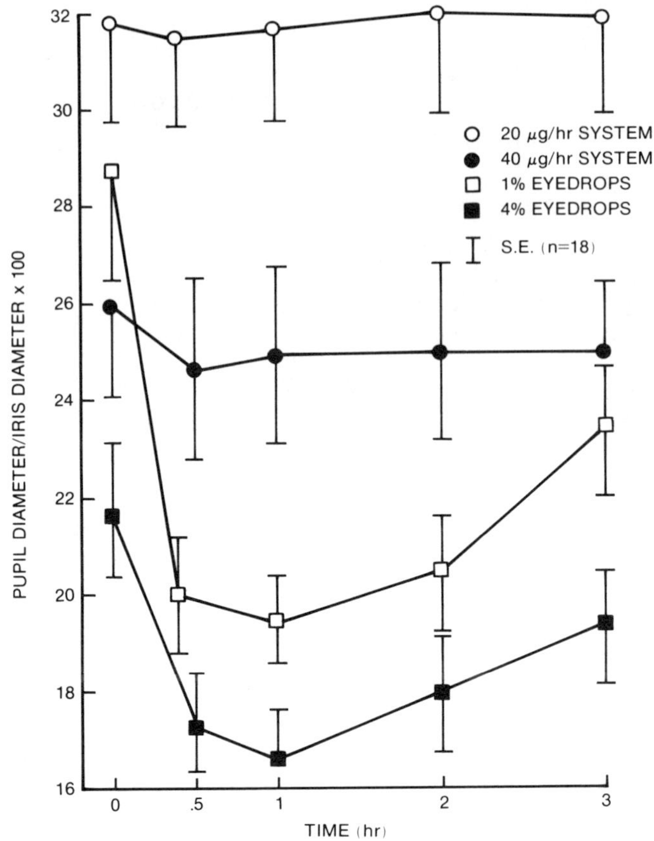

Figure 15. Average miosis of glaucoma patients after receiving pilocarpine in 1% eyedrops, 4% eyedrops, Ocusert® Pilo-20, and Ocusert® Pilo-40. (From Brown, H. S., Meltzer, G., Merrill, R. C., Fisher, M., Ferré, C., and Place, V. A., *Arch. Ophthalmol.*, 94, 1716, 1976, American Medical Association. With permission.)

that the system would promote greater patient compliance. Studies indicate that 30 to 60% of patients fail to take as directed the necessary number of eyedrops per day over 1 week.[29,30] Treating an asymptomatic disease with medications that cause side effects and require frequent administration promotes noncompliance. Such noncompliance may be the most serious difficulty that clinicians face in trying to manage intraocular pressure levels. After using the Ocusert® system for 3 months, 91% of 139 patients preferred it to eyedrops.[31]

CONCLUSION

This article reviewed two products, based on the technology of rate-controlled drug delivery, that reduce toxicity and target the delivery of drug.

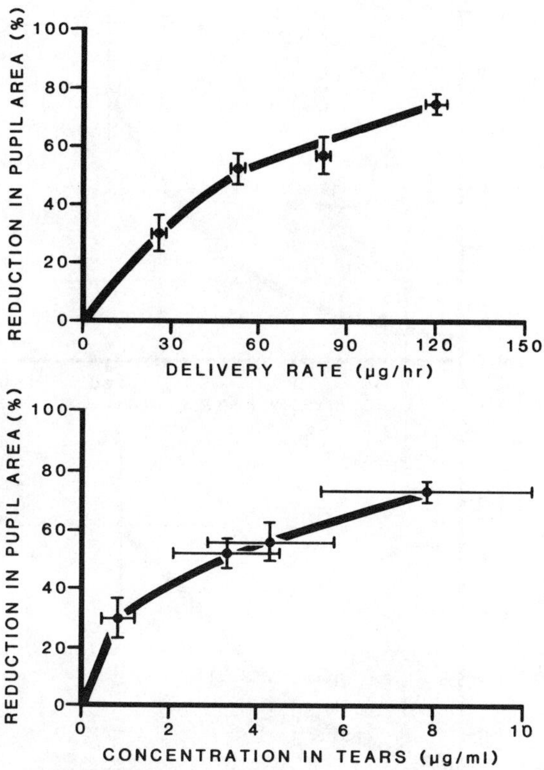

Figure 16. Correlation of miotic effect of pilocarpine with pilocarpine delivery rate and concentration in tears. Pupil area prior to drug treatment served as the basis for calculating the percentage reduction in area during treatment. For each of five subjects, Ocusert® systems were worn for approximately 18 h before and during measurements. (Reprinted with permission from Case Study 2: Development of the Ocusert® Pilocarpine Ocular Therapeutic Systems — A Case History in Ophthalmic Product Development, *Ophthalmic Drug Delivery Systems,* pp. 105–118, copyright 1980, by the American Pharmaceutical Association, 2215 Constitution Ave., NW, Washington, DC 20037, (202)628-4410.)

In this field of controlled drug delivery it is expected that further work with current technologies and development of new technologies will continue to reduce drug toxicity, increase drug efficacy, improve targeted drug delivery, and promote patient compliance.

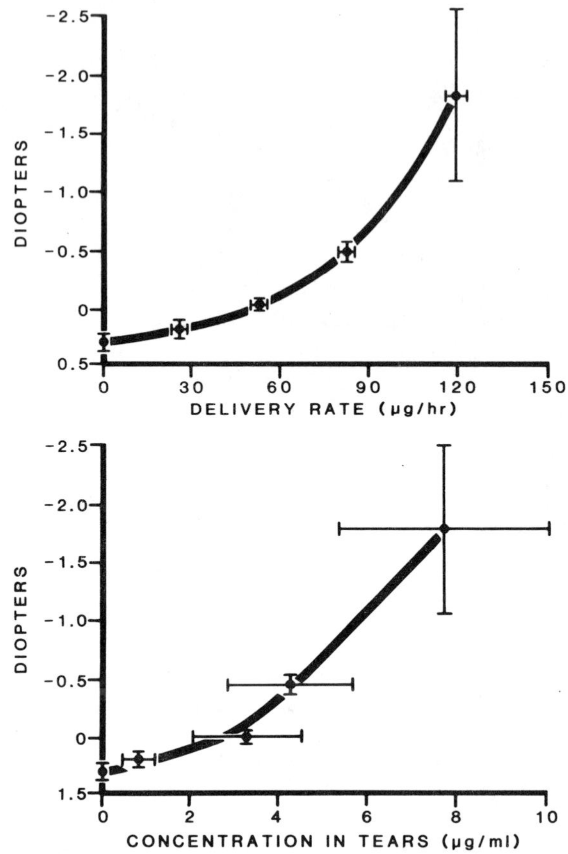

Figure 17. Correlation of induced myopic effect of pilocarpine with pilocarpine delivery rate and concentrations in tears. These measurements were made concomitantly with those shown in Figure 16 on the same subjects. (Reprinted with permission from Case Study 2: Development of the Ocusert® Pilocarpine Ocular Therapeutic Systems — A Case History in Ophthalmic Product Development, *Ophthalmic Drug Delivery Systems,* pp. 105–118, copyright 1980, by the American Pharmaceutical Association, 2215 Constitution Ave., NW, Washington, DC 20037, (202)628-4410.)

TABLE 8
Average Intraocular Pressures of 18 Glaucoma Patients Receiving Pilocarpine by Eyedrop or by the Ocusert® System

	Time (h)[a]				
Drug regimen	0	1/2	1	2	3
1% drops	20.75	19.50	18.55	18.55	18.50
4% drops	19.65	18.45	17.70	17.80	17.80
20 μg/h system	20.10	19.85	19.35	18.70	19.05
40 μg/h system	19.30	19.20	18.85	18.85	19.10
Off medication[b]	28.35	—	—	—	—

Note: Each score is the average for both eyes.

[a] For patients wearing 20-μg/h or 40-μg/h systems, the zero hour is 45 to 48 h after insertion of the system. Measurements were made at the same time on all patients. The "zero" value for patients receiving eyedrops is 6 h after the previous eyedrop, and the other measurements are 1/2, 1, 2, and 3 h after receiving the present eyedrop.

[b] Averages for all patients' intraocular pressures at the time of diagnosis, before treatment or when patients' previous medication was discontinued after treatment was started.

Modified from Brown, H. S., Meltzer, G., Merrill, R. C., Fisher, M., Ferré, C., and Place, V. A., *Arch. Ophthalmol.*, 94, 1716, 1976.

TABLE 9
Contrasts Between the Ocusert® System and Eyedrop Therapy

	Eyedrops 2%, qid	Ocusert® Pilo-20	Ocusert® Pilo-40
Amount of pilocarpine delivered (7 d), mg	28	3.4	6.7
Number of applications	28	1	1

REFERENCES

1. Anon., Towards a new concept of precision in drug administration, *Pharm. J.*, 205, 400, 1970.
2. Zaffaroni, A.: *New Approaches to Drug Administration,* 31st Int. Congress Pharmaceutical Sciences, Washington, D.C., 19, 1971.
3. Urquhart, J., Introduction to Drug Delivery Systems, Proc. Int. Conf., Special Issue of *Pharm. Technol.*, 10, 1983.
4. Pharriss, B. B., Erickson, R. E., Bashaw, J., Hoff, S., Place, V. A., and Zaffaroni, A., PROGESTASERT: a uterine therapeutic system for long-term contraception. I. Philosophy and clinical efficacy, *Fertil. Steril.*, 25, 915, 1974.

5. Urquhart, J., *Novel Methods of Ocular Drug Delivery,* Proc. 6th Int. Congress of Pharmacol. 5. IUPHAR, Helsinki, 20 to 25 July 1975.
6. ALZA Brochure, *Noncompliance with Miotic Therapy: the Views of Four Leading Clinicians,* ALZA Corporation, Palo Alto, CA, 1977.
7. ALZA Monograph, *The PROGESTASERT® Intrauterine Progesterone Contraceptive System,* ALZA Corporation, Palo Alto, CA, 1976.
8. Tillson, S. A., Leong, P., Hudson, B., and Swisher, D., *Preclinical evaluation of intrauterine progesterone contraceptive system,* in Symp. Proc. Clin. Exp. with the Progesterone Uterine Therapeutic System, Acapulco, 15 to 16 October 1976; *Excerpta Medica,* 18, 1978.
9. Leone, U., Uterine progesterone system (UPS): a new form of intrauterine contraception — preliminary report, *Int. J. Fertil.,* 19, 17, 1974.
10. Martinez-Manautou, J., Maqueo, M., Aznar, R., Pharriss, B. B., and Zaffaroni, A., Endometrial morphology in women exposed to uterine systems releasing progesterone, *Am. J. Obstet. Gynecol.,* 121, 175, 1975.
11. Tillson, S. A., Marian, M., Hudson, R., Wong, P., Pharris, B., Aznar, R., and Martinez-Manautou, J., The effect of intrauterine progesterone on the hypothalamic-hypophyseal-ovarian axis in humans, *Contraception,* 11, 179, 1975.
12. Brenner, P. F., Cooper, D. L., and Mishell, D. R., Clinical study of a progesterone-releasing intrauterine contraceptive device, *Am. J. Obstet. Gynecol.,* 121, 704, 1975.
13. Cole, H. M., Questions and answers: diagnostic and therapeutic technology assessment (DATTA), *JAMA,* 261, 2127, 1989.
14. Ginsburg, K. A. and Moghissi, K. S., Alternate delivery systems for contraceptive progestins, *Fertil. Steril.,* 49 (Suppl. 5), 16S, 1988.
15. Machol, L., A fresh look at barrier contraceptives, *Contemp. Ob./Gyn.,* 132, March 1988.
16. Lin, T. J., Billiar, R. B., and Little, B., Metabolic clearance rate of progesterone in the menstrual cycle, *J. Clin. Endocrinol. Metab.,* 35, 879, 1972.
17. Ramos, R. A., Lozano, M., and Martinez-Manautou, J., Contraception with progesterone-releasing intrauterine systems, *Ginecol. Obstet. Mex.,* 37, 69, 1975.
18. Nilsson, C. G., Lahteenmaki, P. L. A., and Luukkainen, T., Ovarian function in amenorrheic and menstruating users of a levonorgestrel-releasing intrauterine device, *Fertil. Steril.,* 41, 52, 1984.
19. Ermini, M., Carpino, F., Petrozza, V., and Benagiano, G., Distribution and effect on the endometirum of progesterone released from a Progestasert device. *Hum. Reprod.,* 4, 221, 1989.
20. Ermini, M., Carenza, L., Pala, A., and Benagiano, G., Effetto dei dispositivi intrauterini liberanti progesterone sull'asse ipotalamo-ipofiso-ovarico nella donna, *Patol. Clin. Ostet. Ginecol.,* 4, 109, 1976.
21. Hagenfeldt, K., Landgren, B. M., Edstrom, K., and Johannisson, E., Biochemical and morphological changes in the human endometrium induced by the Progestasert device, *Contraception,* 16, 183, 1977.
22. Wan, L. S., Hsu, Y. C., Ganguly, M., and Bigelow, B., Effects of Progestasert on the menstrual pattern, ovarian steroids and the endometrium, *Contraception,* 16, 417, 1977.
23. Place, V. A., Fisher, M., Herbst, S., Gordon, L., and Merrill, R. C., Comparative pharmacologic effects of pilocarpine administered to normal subjects by eyedrops or by ocular therapeutic systems, *Am. J. Ophthalmol.,* 80, 706, 1975.

24. Mitra, A. K., Ophthalmic drug delivery device, *Drug Delivery Devices: Fundamentals and Applications,* Tyle, P., Ed., Marcel Dekker, New York, 1988, 456.
25. Urquhart, J., Development of the OCUSERT® pilocarpine ocular therapeutic systems — a case history in ophthalmic product development, in *Ophthalmic Drug Delivery Systems,* Robinson, J. R., Ed., American Pharmaceutical Association, Academy of Pharmaceutical Science, Washington, D.C., 1980, 105.
26. Brown, H. S., Meltzer, G., Merrill, R. C., Fisher, M., Ferré, C., and Place, V. A., Visual effects of pilocarpine in glaucoma: comparative study of administration by eyedrops or by ocular therapeutic systems, *Arch. Ophthalmol.,* 94, 1716, 1976.
27. Heilmann, K., *Long-Term Follow-Up of Controlled-Release Pilocarpine Therapy,* 2nd Int. Conf. Drug Absorption, Edinburgh, 21 to 23 September 1983.
28. Lerman, S. and Reininger, B., Simulated sustained release pilocarpine therapy and aqueous humor dynamics, *Can. J. Ophthalmol.,* 6, 14, 1971.
29. Vincent, P. A., Patient's viewpoint of glaucoma therapy, *The Sight Saving Rev.,* 42, 213, 1972.
30. Spaeth, G. L., Visual loss in a glaucoma clinic. I. Sociological considerations, *Invest. Ophthalmol.,* 9, 73, 1970.
31. ALZA monograph, *The OCUSERT® (Pilocarpine) Pilo-20/Pilo-40 Ocular Therapeutic System: a New Approach to Treating Ocular Hypertension,* ALZA Corporation, Palo Alto, CA, 1974, 49.

Chapter

7

Delivery Systems for Reducing Drug Toxicity: Redox Carriers for Penetrating the Blood-Brain Barrier and Molecusol™ Brand Cyclodextrin for Drug Solubilization

Warren C. Stern
Cato Research
Durham, NC

INTRODUCTION

Ideally, therapeutic agents should be concentrated in the target organ and should be absent or minimally present elsewhere. The absence of the drug outside of the intended site of action would serve to eliminate most adverse reactions. Conceptually, such an ideal biodistribution can be achieved in several ways:

A. Combining the drug with an antibody which is highly selective to the target organ. For the most part this approach has yet to work effectively, since release of a drug from the antibody is problematic, at best.
B. Direct injection of the drug into the organ of interest. In most diseases this approach is impractical and may not provide full perfusion of the organ with the therapeutic agent.

0-8493-8896-1/93/$0.00+$.50

C. Enzymatic activation of the drug by cellular processes which occur in the desired organ and not elsewhere. This approach has been successfully achieved for certain prodrugs, especially antivirals which are phosphorylated mainly by virally induced enzymes. Compounds such as acyclovir have therefore an excellent safety margin since it is activated only in herpes virus-infected cells.
D. Use of a chemical-delivery system in which a carrier molecule processing special biochemical properties achieves selective concentration of the drug in the target organ (see below).
E. Use of special vehicles which minimize formulation-related side effects, e.g., avoiding large fluid loads, irritation upon injection, etc. (see below).

This paper describes drug-delivery systems which act in manners D and E, research in which our scientists and university collaborators have been engaged for the past 5 to 10 years.

BRAIN DELIVERY SYSTEM

The brain, spinal cord, and retina of the eye are largely protected from toxins, infectious organisms, and other foreign molecules by a special vascular barrier, the blood-brain barrier (BBB). The BBB is an endothelial cell barrier in which the endothelial cells which comprise the blood vessels of the central nervous system (CNS) are tightly packed together. By contrast, in other organs there are gaps between the adjacent endothelial cells which permit water-soluble substances to freely move from the blood into the surrounding tissue. The net consequence of this protective vascular barrier is that drugs intended to treat nervous system disorders need to be given in sufficiently high amounts to overcome the adverse concentration gradient experienced between plasma and brain. The main exception to this situation occurs with highly lipophilic drugs, drugs with sufficient fat solubility to readily diffuse through the phospholipid membranes of the capillary endothelium in the CNS, thereby penetrating the BBB.

Redox Carrier System for Delivery of Drugs to the Brain

The work described in this paper regarding redox carriers for delivery of drugs to the brain comes primarily from the pioneering work of Nicholas Bodor at the University of Florida.

It has been recognized for some time that molecular modifications of drugs to enhance lipophilicity will increase brain penetration. However, such drugs will usually maintain an equilibrium between brain and other tissues which parallels over time the elimination kinetics from the body.

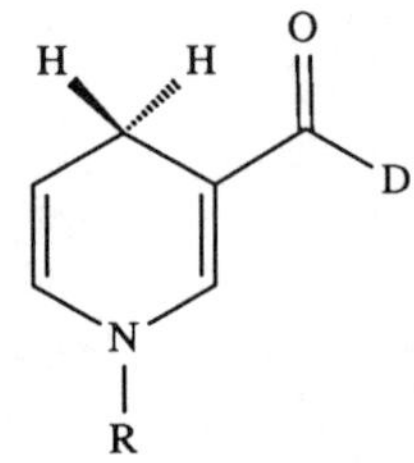

Figure 1. Dihydropyridine carrier connected to a drug (D).

The redox carrier approach (Bodor, 1987), which builds upon the importance of lipophilicity, adds other advantages from the side effect standpoint — reduction in the amount of free drug outside of the brain, blunting of peak levels of free drug in the CNS, and sustained release of free drug in the brain.

The primary carrier molecule employed to date is 1,4-dihydropyridine (DHP, Figure 1). DHP is structurally similar to trigonelline, a natural constituent of coffee beans and bananas, and is attached by a chemical bond to a site on the drug of interest. The DHP drug is now 10 to 100,000 times more lipophilic, thereby enhancing penetration of the BBB. The DHP-drug combination furthermore is usually biologically inactive. After dosing with the carrier-drug, the combination is distributed throughout the body (including the CNS). Then, the DHP moiety is oxidized by the NAD-NADH enzyme system found in all living cells. This converts the redox carrier-drug to a charged pyridinium state, DHP^+-drug. This produces up to a 100,000-fold increase in hydrophilicity for the combination molecule, thereby trapping it in the CNS behind the BBB. In the periphery the quaternary DHP^+-drug molecule is highly water soluble, and as a charged molecule is rapidly excreted by the kidney. Thus, over a 4- to 12-h period DHP^+-drug is cleared from the body, leaving DHP^+-drug trapped in the brain. Concurrently, nonspecific esterases which are ubiquitous slowly cleave the DHP^+ from the drug, giving a sustained release of active compound in the brain. This gradual release of free drug from the carrier-drug combination avoids the sharp peak of free drug levels associated with typical lipophilic compounds and provides a sustained release such that dosing frequency is once per day or less often. The DHP^+ itself, a small-molecular weight quaternary compound, is actively excreted from the brain and eliminated in the urine. Eventually, the free drug liberated from the carrier-drug in the CNS leaks out via the blood or CSF system, and 12 to 48 h after dosing another dose is needed.

The net effect of the redox carrier system, in addition to enhancing efficacy as a result of higher drug concentrations in the CNS, is to reduce the systemic toxicity usually associated with the drug being delivered. The reduction in toxicity occurs due to less free drug in the rest of the body and

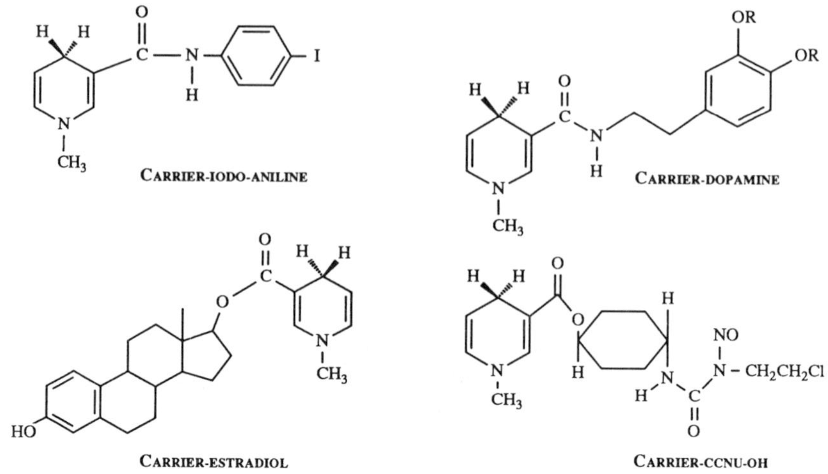

Figure 2. Chemical structure of several carrier-drug combinations.

to a lower total dose of drug that needs to be administered in order to achieve the desired therapeutic levels in the brain. Examples of reduced body load of carrier-drug in the presence of higher brain levels comes from studies of an iodinated imaging agent, carrier-dopamine, carrier-estradiol, and carrier-CCNU-OH (alkylating chemotherapeutic). The structures of these DHP-linked moieties are shown in Figure 2.

Iodine

Studies in rodents by Tedjamulia et al. (1985) showed the i.v. dosing of I^{125}-aniline connected to the DHP carrier produced *blood* levels 5 to 60 min post-dose of 0.31 to 0.37% dose per gram. This compared to 0.54 to 0.94% dose per gram for injection of an equimolar amount of I^{125}-aniline (without the carrier). Levels of radioactivity in *liver* at 1 h post-dose were also reduced by about 50% using the DHP system compared to the situation without the carrier (0.82% of injected dose vs. 1.60% dose per gram). In contrast, *brain* levels were about 20 times higher using the carrier system than without. Thus, on an overall ratio of brain:blood or brain:liver, radioactivity levels were improved 40-fold by use of the DHP approach. This leads to the prediction of enhanced imaging in the brain while reducing the radioactivity exposure of peripheral organs.

Dopamine

Dopamine, a sympathetic neurotransmitter, usually does not enter the brain from the periphery, and in the periphery exerts profound sympathetic effects on the cardiovascular system. I.v. infusion of 50 mg/kg of dopamine connected to the carrier, which produced 10 to 25 μg/g concentration of

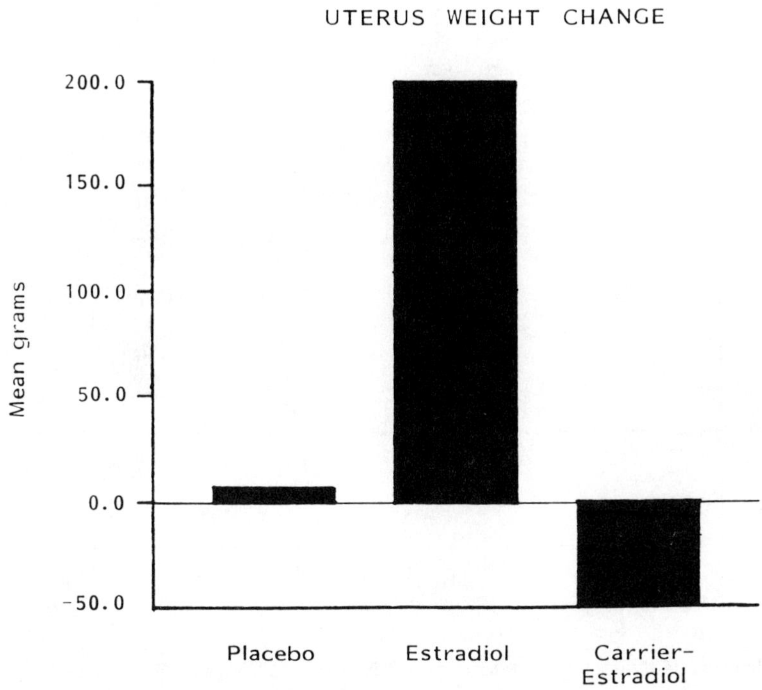

Figure 3. Uterine weights in rats following i.v. injection of 3 mg/kg carrier-estradiol or equimolar estradiol.

carrier-dopamine in the brain (Bodor and Simpkins, 1983) did not produce cardiovascular collapse, whereas lower i.v. doses of doapmine alone in rats is lethal due to sympathetic excess.

Estradiol

This female sex hormone is well known to exert peripheral estrogenic effects on the uterus, i.e., uterine hypertrophy. I.v. dosing of intact female rats with 3 mg/kg of the estradiol carrier combination, or an equimolar amount of 17-β-estradiol produced dramatically different effects on uterine weight (Figure 3). While brain estradiol levels were enhanced many times over that seen with estradiol alone (Bodor et al., 1987), there was no uterine hypertrophy in intact rats after administration of the DHP-estradiol combination (Sarkar et al., 1987).

CCNU-OH

I.v. injections of the redox carrier combined with CCNU-OH produced in rats a 10- to 50-fold enhancement in brain levels of the chemotherapeutic

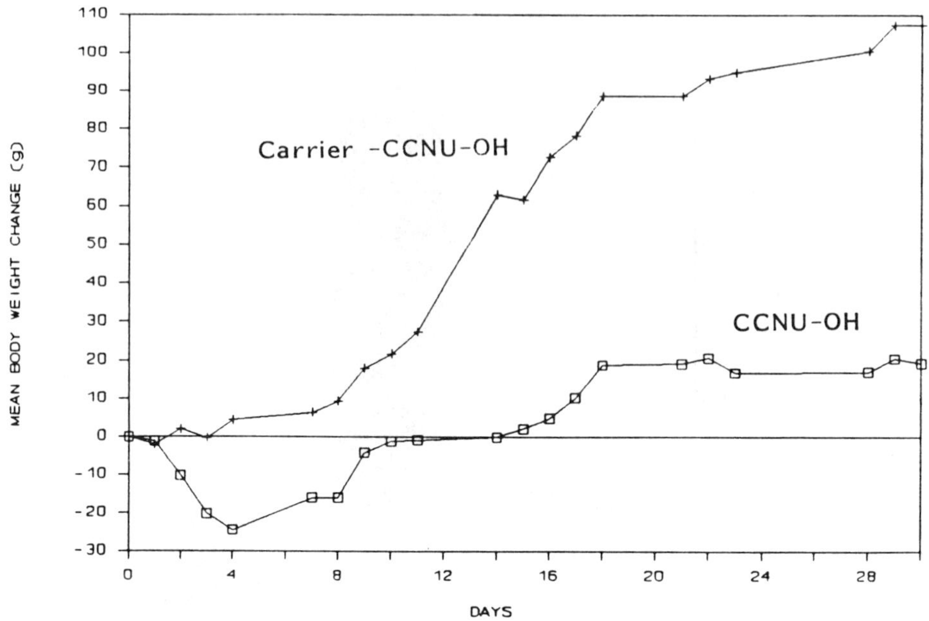

Figure 4. Body weight gain after carrier-CCNU-OH or CCNU-OH, i.v. dosed on day 0.

agent over that seen with CCNU-OH itself for at least the first hour after dosing (Raghavan et al., 1987). Side effects of this alkylating agent monitored over the next 30 d following a single 50-mg/kg dose showed fewer deaths and much less body weight loss with the redox carrier system than without (Figure 4).

The basis for reduction in adverse drug reactions in tissues outside the CNS when using the redox carrier system is several-fold — a reduction in circulating levels of free drug, the presence largely of quaternary-drug complexes outside the brain (which are usually biologically inactive), and rapid elimination of the quaternary-drug via the kidney. Additionally, the total milligrams of carrier-drug to be administered in order to achieve a desired therapeutic concentration in the brain is often 1/5 to 1/50 of the standard dose, further reducing the opportunity for side effects in other organs.

Clearly, there is a risk for greater CNS toxicity when a delivery system causes the brain to see more free drug than has been achieved before. The potential for enhanced CNS toxicity therefore must be carefully evaluated in animal toxicity studies and in risk benefit judgments during human therapeutic testing. One aspect, however, of the DHP delivery system is that neurotoxicity similar to that reported with MPTP simply does not occur (Brewster et al., 1988). In general, the redox carrier system, when combined with a drug, seems to make the drug better tolerated from the standpoint of CNS and other organs.

CHEMICALLY MODIFIED CYCLODEXTRINS

A different type of drug toxicity can be attenuated by drug-delivery systems which enable the creation of improved formulations. Historically, the value of enteric coatings in reducing stomach irritation to ulcerogenic compounds is a well-established precedent for the importance of formulation characteristics in attenuating side effects of drugs.

Adverse reactions associated with the formulations of parenteral products of several types — pain upon injection, phlebitis, excessive fluid loads — are all caused by irritating organic solvents, abnormal pH, or large volumes of water employed due to poorly soluble drugs. Importantly, extravasation reactions, i.e., tissue necrosis which occurs upon extravascular injection of cytotoxic drugs or other highly chemically reactive substances, is a significant local site toxicity which is amenable to improvement by better formulations.

For the past 3 years we have been working with a novel solubilizing system, hydroxypropyl-cyclodextrin (HPCD, sold under the tradename Molecusol® by Pharmatec, Inc.), an excipient which obviates many of the common injection site toxicities. We came to appreciate the important clinical properties of HPCD in the course of resolving the solubility and stability problems characteristic of the lipophilic and oxidatively susceptible redox carrier drugs. After working with HPCD and redox-linked drugs it became apparent that there was a strong rationale for use of HPCD as a general delivery system for injectable drugs and other dosing formulations.

Description of the Technology

HPCD originates from the field of cyclodextrin research. Cyclodextrins, of which there are many types, are starches produced by bacterial or enzymatic processes using natural carbohydrates as a substrate. The chemical structure, synthesis, analytic methods, and general physical chemical properties of cyclodextrins are well known and the field of research is 100 years old. Currently, β-cyclodextrins are used in the food industry abroad to preserve flavorings, in industrial synthesis processes to modify reactions, and to a small extent in pharmaceutical formulations (e.g., to stabilize prostaglandins) in Europe and Japan (see Szejtli, 1988). For a variety of reasons related to limited solubility, limited patent coverage, and questions of safety and cost, the applications of the natural cyclodextrins in pharmaceutical products have been quite limited. In the U.S., no drug products currently employ cyclodextrin technology. We believe that due to recent advances in novel types of chemically modified cyclodextrins, specifically HPCD, many important uses will soon occur with formulations intended for all routes of administration for therapeutics and diagnostic agents.

HPCD is a chemically modified β-cyclodextrin which has dramatically greater solubility than other cyclodextrins and which, based on animal toxi-

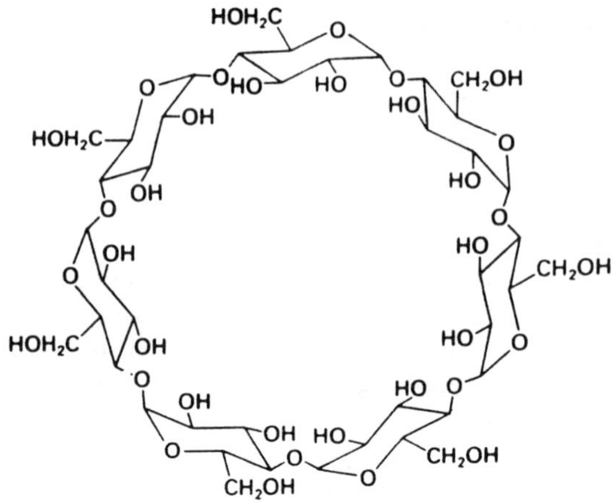

Figure 5. Basic ring structure of a β-cyclodextrin.

cology studies, should be safe for pharmaceutical use (Pitha, 1988; Brewster et al., 1988). As shown in Figure 5, the general structure of a cyclodextrin is a circular molecule with a hollow center. The outer edge of the HPCD molecule is highly water soluble, the inner core is highly fat soluble. This permits HPCD to form reversible "inclusion complexes" with drugs which are poorly water soluble — one molecule of the drug fits into the center of one molecule of HPCD. The HPCD:drug complex can then be solubilized in water at neutral pH (or under acidic or basic conditions if needed) with the inclusion complex remaining intact as long as adequate concentrations of cyclodextrin:drug are maintained. Once this concentration falls below a critical level unique to each drug, the inclusion complex very rapidly falls apart. This occurs when the HPCD:drug is administered to the stomach or the blood, i.e., free drug is released in its original form in a larger volume of fluid. Typically, 1 g of HPCD can solubilize 10 to 200 mg of a drug, although higher or lower values are sometimes achieved.

Examples of Reduction in Drug Toxicity by HPCD

A. Elimination of Abnormal pH

Because of limited aqueous solubility at neutral pH, certain commonly used drugs must be given in a highly basic or acidic formulation. For example, i.v. thiopental has a pH of 10 to 11 and injectable chlordiazepoxide has pH 3. Both compounds are solubilized readily by HPCD at neutral pH (Table 1) and at milligrams per milliliter concentration at least equal to the current basic formulations.

TABLE 1
Reduction in Toxicity Resulting from Improved Parenteral Formulations Using HPCD

Drug	Therapeutic class	Current formulation		Maximum solubility aqueous HPCD formulation[a] mg/ml
		Vehicle	mg/ml	
Thiopental	General anesthetic	Water (pH 10–11)	2–50	106
Chlordiazepoxide	Antianxiety, sedative	Tween® 80 (pH 3)	50	109
Diazepam	Antianxiety, sedative	10% ethanol	5	5
Digoxin	Cardiotonic	10% ethanol	0.25	52
Vitamin E	Nutrient	Cremophor[a]	50 U/ml	>50 U/ml
Adriamycin	Anticancer	Water	2	5
Dexamethasone	Anti-inflammatory	Water[c]	4–24	50

[a] Neutral pH.
[b] Replaced by oil/propylene glycol multivitamin mixture.
[c] Phosphate salt.

B. Undesirable Organic Solvents

Injectable digoxin is solubilized in 10% ethanol, an irritating solution. Vitamin E injectable products (E-ferol) had been given in Cremophor, a castor oil solvent which produced anaphylactic reactions and death in a substantial number of infants who received this formulation. Both compounds are readily soluble in aqueous solutions of HPCD (Table 1).

C. Extravasation Due to Chemical Reactivity of the Drug

Chemotherapeutic agents are commonly recognized for their ability to induce venous irritation and, if given extravascularly, to produce painful and prolonged tissue necrosis (Dorr, 1981). Complexation of adriamycin with HPCD was found to reduce extravasation reactions by 30% in a rodent model involving intradermal injection (Simpkins, unpublished observation). The basis for this decrease is that when a drug is complexed with HPCD it is much less chemically reactive. Only upon dissociation of the drug from the complex does the free drug emerge to react with tissues. Such dissociation occurs only after substantial dilution of the complex, thereby diluting the concentration of free drug in comparison to what would have occurred initially if the drug were given without HPCD.

D. Other Benefits

Anti-inflammatory steroids are highly insoluble in water. The pharmaceutical industry has employed water-soluble prodrugs, ester salts which are cleaved by ubiquitous body enzymes. In the case of dexamethasone PO_3, the salt is associated with diffuse burning and itching sensations. Use of this ester can be obviated by HPCD, since aqueous solutions of dexamethasone base in HPCD are readily achieved (Table 1).

The biodistribution within the body of drugs injected in an HPCD solution rather than an organic solvent (DMSO) can also be improved. For example, in rodent studies we have seen much lower lung levels of an experimental steroid when it is dissolved in HPCD rather than DMSO (unpublished observations). This probably reflects a reduction in precipitation of the lipophilic drug in plasma which occurs when a drug in an organic solvent solution is injected into the aqueous bloodstream — the lung tends to trap fine particles of precipitated drug.

Lastly, cyclodextrins can reduce side effects of orally administered medications by masking of taste of solutions and by reducing gastric irritation (Szejtli, 1988).

In summary, HPCD is a new excipient which is currently being evaluated for use in formulations by various companies. Although its safety is not fully established due to its newness, we believe it offers excellent promise for reducing drug toxicity for many classes of drugs and especially for the parenteral route of administration.

REFERENCES

Bodor, N., Redox drug delivery systems for targeting drugs to the brain, *N.Y. Acad. Sci.,* 507, 289, 1987.

Bodor, N., J. McCornack, and M. E. Brewster, Improved delivery through biological membranes. XXII. Synthesis and distribution of brain-selective estrogen delivery sytems, *Int. J. Pharm.,* 35, 47, 1987.

Bodor, N. and J. Simpkins, Redox delivery system for brain-specific, sustained release of dopamine, *Science,* 221, 65, 1983.

Brewster, M. E., K. S. Estes, R. Perchalski, and N. Bodor, A dihydropyridine conjugate which generates high and sustained levels of the corresponding pyridinium salt in the brain does not exhibit neurotoxicity in the monkey, *Neurosci. Lett.,* 87, 277, 1988.

Brewster, M. E., K. S. Estes, J. F. Howes, and N. Bodor, Acute and subacute toxicity of 2-hydroxypropyl-β-cyclodextrin in Sprague-Dawley rats and cynomolgus monkeys, presented at Fifth International Symposium on Inclusion Phenomena and Molecular Recognition, Orange Beach, AL, September 1988.

Dorr, R. T., Extravasation of vessicant antineoplastics: clinical and experimental findings, *Arizone Medicine,* April 1981, pp. 271–274.

Pitha, J., Pharmaceutical preparations containing cyclodextrin derivatives, *U.S. Patent* 4727064, February 1988.

Raghavan, K. S., E. Shek, and N. Bodor, Improved delivery through biological membranes. XXX. Synthesis and biological aspects of a 1,4-dihydropyridine based chemical delivery system for brain-sustained delivery of hydroxy CCNU, *Anti-Cancer Drug Design,* 2, 25, 1987.

Sarkar, D. K., S. J. Friedman, S. S. Yen, and R. Roy, Brain-directed delivery of estradiol in rats, presented at Society for Neuroscience, November 1987, New Orleans.

Szejtli, J., *Cyclodextrin Technology,* Kluwer Academic Publishers, Dordrecht, The Netherlands 1988.

Tedjamulia, M. L., P. C. Srivastava, and F. F. Knapp, Jr., Evaluation of the brain-specific delivery of radioiodinated (iodophenyl)-alkyl-substituted amines coupled to a dihydropyridine carrier, *J. Med. Chem.,* 28, 1574, 1985.

Chapter

8

Example of the Use of Pharmacokinetic Modeling to Compare Relative Risks of Different Exposure Scenarios*

Jerry N. Blancato
Exposure Monitoring Program
Exposure Assessment Research Division
Environmental Monitoring Systems Laboratory
U.S. Environmental Protection Agency
Las Vegas, NV

Physiologically based pharmacokinetic models are not new. In the 1930s, Teorell developed the concept that mass balances could be written which adequately described the dispostion of therapeutics and other compounds within the body. The resultant differential equations, which describe the time course of the disposition of the chemical, usually have no analytic solution. At that time, advanced automated computer technology was not available and thus efficient and practical numerical solution of these equations was not possible. As a result, various simplifying assumptions were made, such as describing the body as a one- or two-compartment vessel instead of using the true anatomical description of the body as a basis for model formulation. This and other simplifying assumptions lead to systems of equations which lend themselves to analytical solution. Such models, today generally referred to as classical models, are, however, not based on true anatomical and physi-

* Note: The views expressed in this paper are the views of the author and do not necessarily represent the policies or views of the U.S. Environmental Protection Agency.

0-8493-8896-1/93/$0.00 + $.50

ologic representations of the body. In addition, the physical constants used to quantify transfer between compartments, from the body, and so on, are actually numerically fit to descriptive data. These models, while adequately describing existing data and which are used for optimizing therapeutic dose, are not useful for predictive purposes. With these models, for example, it is very difficult to predict from data in one species to another, or from one exposure regimen to another. The basic reason for this seeming inadequacy is that the physical constants describing transfer and other processes are numerically derived from existing data. When conditions such as the species or the exposure regimen are changed, more data are necessary to fit a new set of constants.

Figure 1 shows a typical diagrammatic representation of a physiologically based pharmacokinetic (PBPK) model which describes disposition into several body organs. QA is the arterial blood flow to the various organs, V is the volume of the organs, and QvR is the venous return from the organs. This model assumed each organ to be a well-stirred compartment. At any time, the instantaneous concentration of a substance in the organ is the difference between the arterial concentration and the concentration in the venous return from the organ.

The concentration of the substance in the systemic venous return is the sum of the concentration of substance in the venous returns from each organ. In actuality, the concentration in the venous return of an organ is related to the actual tissue concentration of the organ in the following manner:

$$\text{Cvi} = \text{Cti/R} \tag{1}$$

Cvi is the concentration in the venous blood of an organ i, Cti is the tissue concentration in an organ i, and R is the partition coefficient of the substance between tissue i and the blood. This particular relationship is a commonly used one, but can be modified when the organ is not well stirred or for conditions where partitioning is concentration dependent. From such a relationship, mass balance equations may be written for each of the organs and for the total body as follows:

$$\text{dCti/dt} = [\text{QAi*(CA} - \text{Cti/Ri)]/Vi} \tag{2}$$

For metabolizing organs such as the liver, the equation may be modified, for example, as follows:

$$\text{dCi/dt} = [\text{QAL*(CA-Cti/RL)} - \text{(Vmax*CL)/(KM} + \text{CL)]/VL} \tag{3}$$

In this particular case, typical Michelis-Menten saturation kinetics is assumed for the liver compartment. The right side of this mass balanced actually means that the concentration of the material in the tissue is equal to what comes in

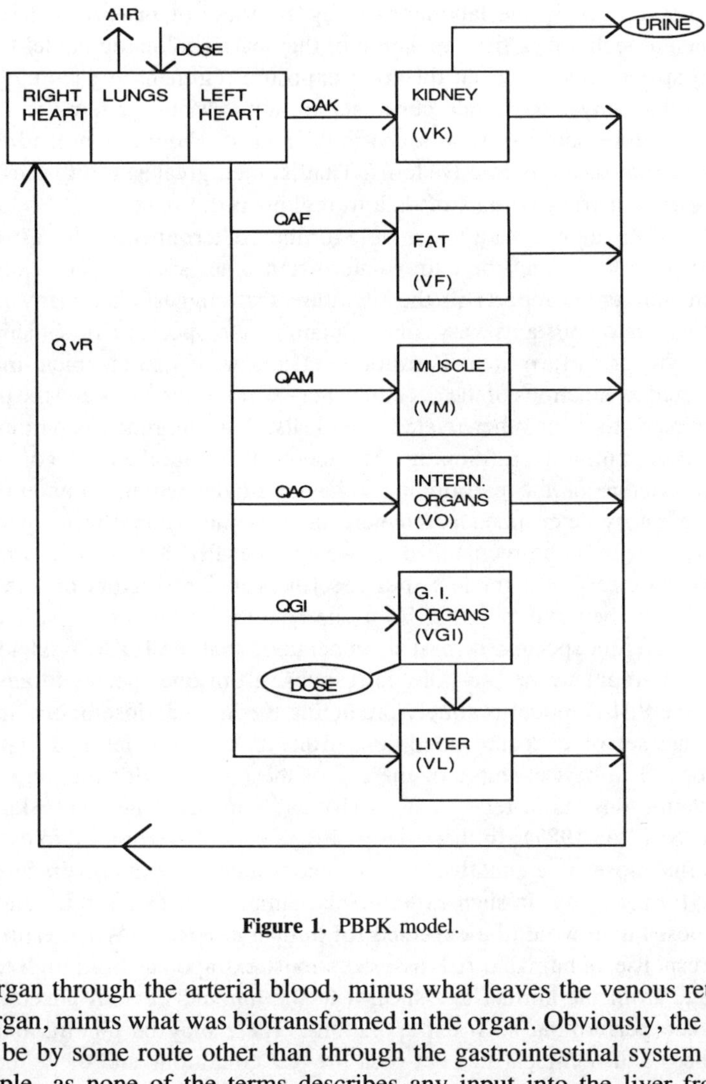

Figure 1. PBPK model.

the organ through the arterial blood, minus what leaves the venous return of the organ, minus what was biotransformed in the organ. Obviously, the dosing must be by some route other than through the gastrointestinal system in this example, as none of the terms describes any input into the liver from the gastrointestinal system.

Thus, it can readily be observed from Figure 1 and from the above equations that all the model parameters have either physiological or physical meaning and can be determined from laboratory experiments. The values for blood flows and organ volumes have been determined in a variety of species and are available in the scientific literature. The thermodynamic parameters such as partition coefficients are determined empirically from known physical properties of chemical and tissue or can be determined in the laboratory. Metabolic rate constants, although sometimes more difficult to estimate, may also be determined in the laboratory using a variety of *in vitro* and *in vivo* experimental techniques. Extrapolation of the pharmacokinetic model for use between species and between different exposure regimens is facilitated because the key physiologic rate constants are appropriately altered to reflect the actual species and exposure scenario of interest. From this it can be seen that PBPK models are predictive tools. That is, their greatest utility is as tools of extrapolation from regions of data to regions of no data.

When data are not readily available for their determination, the estimates of some parameters may be extrapolated from other sources. For example, there are numerous reports in the literature that suggest that many of the physiologic rate constants vary among mammalian species as a function of body weight or surface area. Partition coefficients of one chemical may be extrapolated as functions of the octanol water-partitioning ratios and the tissue-partitioning values for other related chemicals. This approach is particularly useful for determining partitioning coefficients of metabolites of a compound from knowledge of the partitioning behavior of the parent compound and some laboratory-determined parameter, such as the octanol/water partition coefficient. It must be recognized, however, that PBPK models are not alternative dose-response models, but describers and predictors of ''internal dose''. Thus when extrapolating from one condition to another, particularly between different species, it must be understood that the PBPK model is not acting to extrapolate the physiologic response from one species to another. Rather, the PBPK model is simply predicting the internal dose in one species and at one set of exposure conditions from another species and exposure conditions. The best example of the use of these models for the purpose of extrapolating dose is in recent cancer risk assessments (Chen and Blancato, 1987; U.S. EPA, 1987). In these cases, PBPK models were used to estimate the internal dose of a putative carcinogen in animals exposed in bioassay studies. By necessity, in such experiments, animals are exposed to relatively higher doses than would be expected for human exposure. When estimating actual response in humans, risk assessors must extrapolate from high to low doses and from the animal test species to the human. Several uncertainties arise when performing such extrapolations. There may be pharmacokinetic differences and nonlinearities between the test conditions and the actual exposure scenarios. Given differences in macroscopic and microscopic physiology, different species may respond far differently to the same toxin. For example, DNA repair rates, immunosurveillance, number of cells, receptor characteristics and amounts, and levels of intrinsic modulating substances are but a few of the possible inherent differences between species. PBPK models, when properly formulated and validated, may, however, help reduce some of the uncertainties associated with the pharmacokinetics. Thus, a more thorough extrapolation of dose may at least be made between the test and actual conditions of exposure.

An example of a practical use of PBPK models is to compare relative risks from different exposure scenarios. To illustrate their utility in this regard, a model developed to estimate internal doses of the putative toxic species after exposure to tetrachloroethylene is used. The development of the model and its use in the risk assessment process is described elsewhere (Chen and Blancato, 1987), as is the use of the model for selecting the parameters requiring the most detailed laboratory determination (Blancato, 1988). Briefly, the model assumed that the organs of the body are well stirred and perfusion limited. This latter assumption means that the rate of transport into the organs is far greater than the blood-perfusion rate of the material to the organ. For this example, metabolism was assumed to occur only in the liver. The body was modeled as having several compartments including the liver, adipose or fatty tissue, a vessel-rich, and vessel-poor compartment. If sufficient data exist for validation and conditions warrant, the resolution of the model could be expanded to include other organs of interest. The determination of key parameters, including human metabolic rates, has been described (Chen and Blancato, 1987). The objective here is not to further validate or refine the model, but to simply use it for illustrative purposes.

Figure 2 shows model predictions of parent compound concentration in various compartments during and after exposure to 400 ppm tetrachloroethylene for 2 h. The fat compartment, while not necessarily achieving the highest concentrations of parent compound, does, however, sequester the material for longer periods than any other tissue. The model predicts a similar profile for other exposure concentrations. This prediction could be significant with multiple dosing. Under some conditions of multiple dosing, the model shows that the body burden of parent compound is not totally eliminated before the subsequent exposure is started. Therefore, under certain conditions of repeated exposure the body will store the chemical.

The site of putative toxicity for this compound is believed to be the liver. Figure 3 compares model predictions of the concentration of parent compound in the liver after three different exposure conditions. The exposure scenarios used were 100 ppm for 8 h, 400 ppm for 2 h, or 800 ppm for 1 h. All three resulted in an equal mass-time product of 800 ppm-hrs. However, the peak concentrations of parent compound in the liver and the length that the peak concentrations were maintained differs significantly for the three exposure conditions. Depending upon the mechanism of action, if peak concentration value is the dose meter to be used to predict risk, the three exposure scenarios would result in different estimates of risk, with the short 800-ppm exposure being the most dangerous. On the other hand, if the duration of steady-state concentration in the liver of parent compound is judged to also be of concern, then the longer 100-ppm exposure might result in the greatest risk. Interestingly, the impact of the different exposure regimens on the fat concentrations of parent compound is not as dramatic (Figure 4) with all three achieving nearly the same peak concentrations and having similar elimination profiles

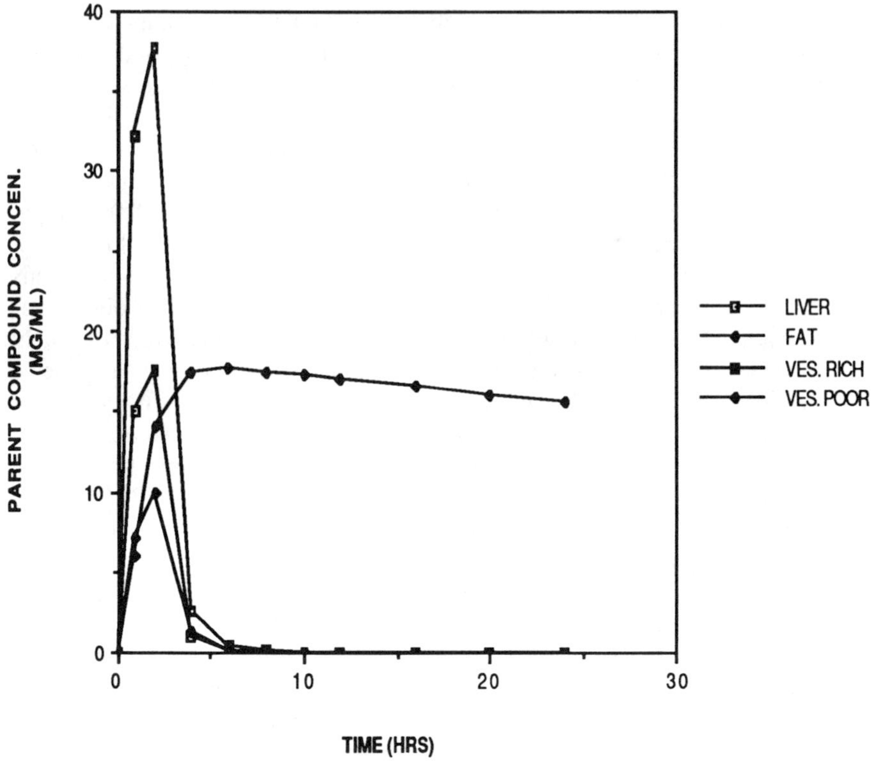

Figure 2. Model results: organ concentrations, 400 ppm exposure, 2-h exposure duration.

once exposure has ceased. The only difference is that with higher exposure concentrations the time to reach steady state is shorter.

Figure 5 compares the model predictions of the area under the concentration curve (AUC, or the mass-time product) for the liver under the three exposure scenarios previously described. For this particular compound the AUC are similar, but the higher exposure concentration results in a greater AUC.

On the other hand, Figure 6 shows model predictions for the amount of parent compound metabolized. In this case the results show that the longer lower concentration exposure produced the greatest amount of metabolite.

Therefore, if risk is dependent upon the total amount of parent present in the liver over time, the 800-ppm, 1-h exposure poses the greatest risk. If, however, risk is dependent upon the amount of metabolite produced, then the 100-ppm, 8-h exposure poses the greatest risk. Remember that both scenarios produced the same 800 ppm-h cumulative exposure.

It is also well to note the sampling of fat for parent compound would not have revealed the subtle differences in the amount metabolized, the AUC in

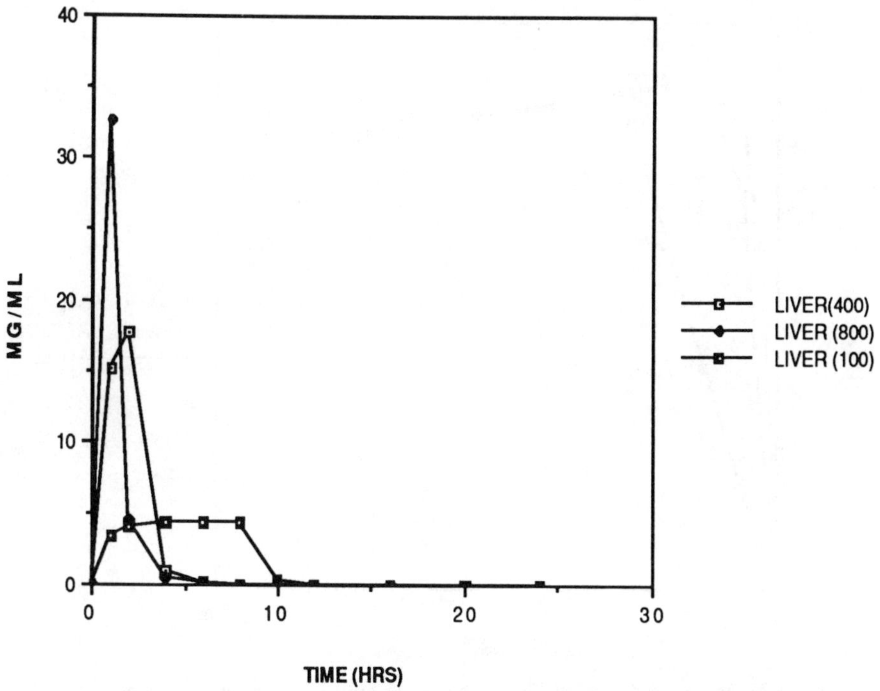

Figure 3. Model results: liver concentrations at various exposure regimens.

the liver, the levels of peak liver concentrations, or the duration of steady-state concentrations that were shown by model predictions.

Before drawing conclusions from results such as these it should be cautioned that like all models, PBPK models need to be first properly formulated and validated. This model was used here for illustrative purposes only and thus no discussion regarding is formulation and validation is presented. These examples discussed illustrate one potentially powerful use of PBPK models. The impact of different exposure scenarios, particularly with repeated exposures, may often be too difficult, time consuming, and potentially dangerous to measure under actual conditions. With the proper validation and formulation, such a model is a useful tool for the exposure and risk assessor for comparing the relative risks that result from different exposure scenarios.

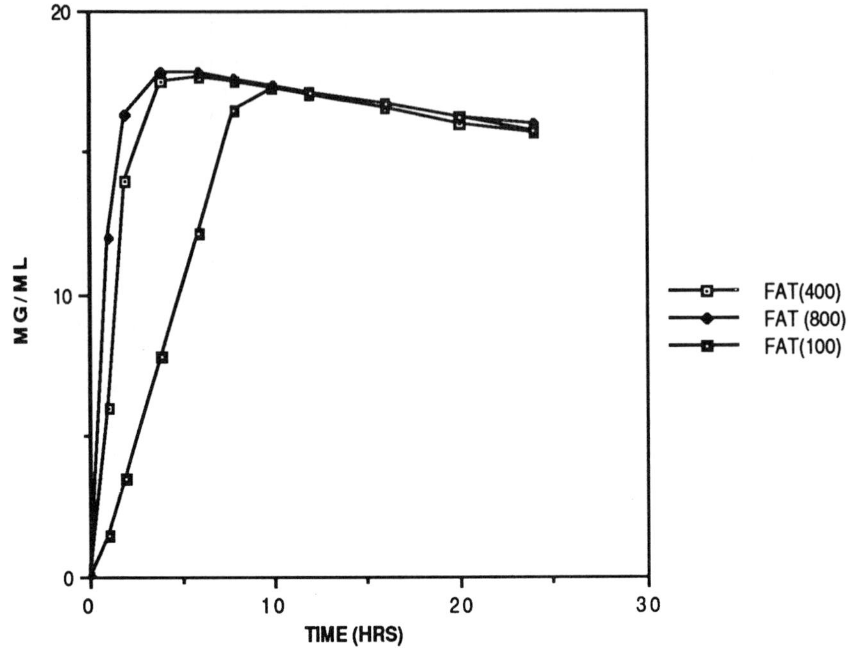

Figure 4. Model results: fat concentrations at various exposure regimens.

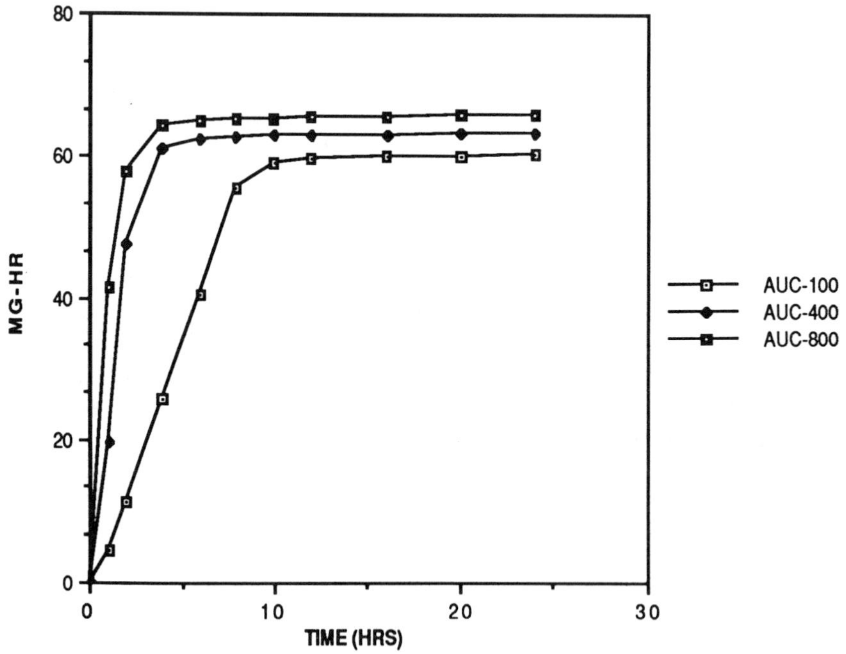

Figure 5. Model results: comparison of area under concentration curve of parent compound in liver for 100-, 400-, and 800-ppm exposure.

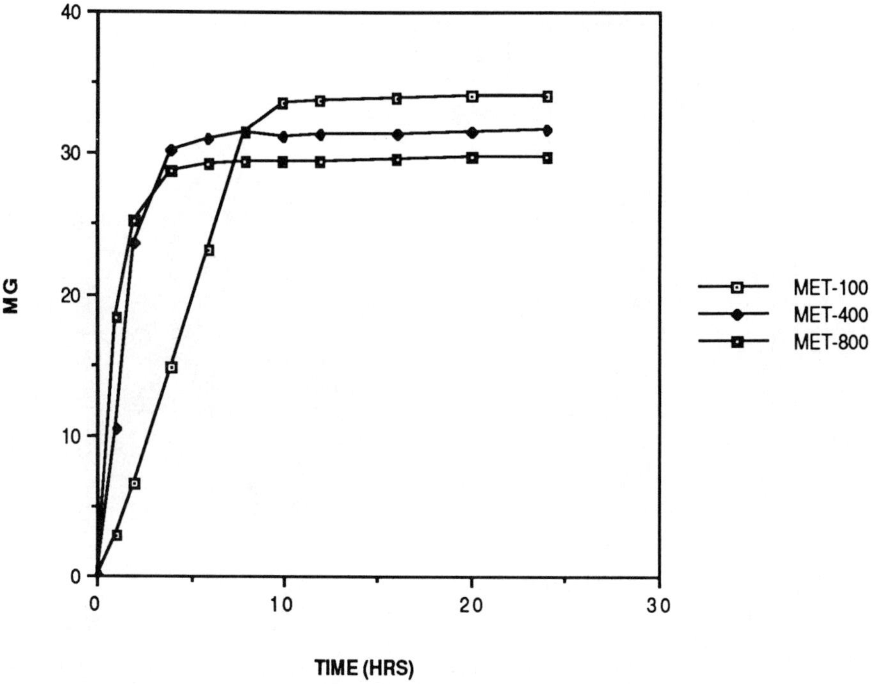

Figure 6. Model results: amount metabolized.

REFERENCES

Blancato, J. N., Use of Simulation Modeling in Exposure and Risk Assessment, NTIS; April 1988; 21p, Report No.: EPA/600/D-88/070.

Chen, C. W. and J. N. Blancato, Role of Pharmacokinetic Modeling in Risk Assessment: Perchloroethylene as an Example, in Pharmacokinetics in Risk Assessment: Drinking Water and Health V.8, National Academy Press, Washington, D.C., 1987.

U.S. EPA, Update to the Health Assessment Document and Addendum for Dichloromethane (Methylene Chloride): Pharmacokinetics, Mechanism of Action, and Epidemiology (Review Draft), EPA/600/8-87/03A, July 1987.

Chapter

9

Application of Radiotelemetry to Cardiovascular Measurements in Pharmacology and Toxicology

Brian P. Brockway
Data Sciences International, St. Paul, MN

Craig R. Hassler
Battelle, Columbus, OH

INTRODUCTION

Measurements from the circulatory system of laboratory animal models play an important role in pharmacologic and toxicologic evaluations, in addition to many biomedical research protocols. For pharmacologists and toxicologists, measurements of parameters such as blood pressure (BP), electrocardiogram (ECG), and heart rate (HR) provide a valuable tool for predicting if a new compound will be safe and effective in human beings.

Traditionally, a variety of techniques have been employed to obtain these measurements. Chemical restraint is probably the simplest method, but the anesthesia employed affects normal body function, and time-course profiles longer than a few hours are difficult to obtain. Consequently, measurements from conscious animals are often preferred. The most common techniques

0-8493-8896-1/93/$0.00 + $.50

for monitoring conscious animals require the use of either a tether or a restraining device while measurements are taken via electrodes, indwelling exteriorized vascular catheters, vascular access ports (VAP), or BP cuffs. The use of restraints and tethers allows data to be collected from awake animals, but can introduce stress artifact in the data and cause animal management problems.

Radiotelemetry provides an alternative means of obtaining measurements from freely moving conscious animals. The scientific literature documents the use of telemetry to monitor laboratory animals as early as the 1950s with much of the pioneering work in telemetry of cardiovascular data being done by Franklin et al.[9] in the 1960s. It should be noted that the early telemetry systems were typically not implantable, but relied upon backpacks attached to the animal and utilized percutaneous catheters where necessary. Today, nonimplantable telemetry plays a very minor role in a few specialized research protocols, but still plays a major role in wildlife research for tracking and long-range monitoring of vital signs. Intuitively, implantable telemetry, in which there are no percutaneous sites, has a number of potential benefits when compared to conventional methods. It can eliminate pathways for infectious agents into the body, provide total freedom for the animal to move about its home cage while measurements are taken, and provide the ability to collect data around the clock without the stress from human contact. One of the first implantable telemetry systems for cardiovascular measurements was described by Deboo and Fryer[10] in 1965. However, it was only recently that the evolution of electronics, packaging, sensor, and battery technology has made it feasible to mass produce reliable, miniaturized implantable devices at a reasonable cost.

REVIEW OF CONVENTIONAL METHODS OF MEASUREMENT

Chemical Restraint

One technique often used for acute studies lasting no more than a few hours is to chemically restrain the animal with anesthesia. However, anesthesia itself produces physiological changes and may affect pharmacodynamics and pharmacokinetics. When collecting data for more than a few minutes, the level of anesthesia may vary, resulting in artifact in the measured response. McGill[55] has pointed out that when measuring BP the use of anesthetics is at a minimum undesirable and in some studies is unacceptable.

The justification for avoiding anesthesia in cardiovascular studies is widely documented in the literature. For instance, Walsh et al.[54] have shown that in rats, ether reduces arterial pressure by about 9%, increases cardiac output by 92%, and reduces total peripheral resistance by 47%. Chloralose-urethane anesthesia is reported to decrease stroke volume, cardiac index, and minute work of the heart[14] and impair reflex responses to tilt and carotid artery occlusion.[15] This anesthetic has also been reported to decrease mean arterial pressure (MAP),[14,16] while HR is either unchanged[14] or decreased.[16] Chloralose-urethane has also been shown to decrease cerebral, gastrointestinal (GI) tract, and myocardial blood flows, but to not affect hepatic or renal blood flow.[16] Many of these effects are strain specific.[14]

Barbiturate anesthetics, pentobarbital, for example, have been shown to reduce respiratory rate, decrease PO_2, and increase PCO_2.[16] Pentobarbital has been shown to decrease cerebral, GI tract, and myocardial flow, but not hepatic or renal blood flow.[16]

Chelly et al.[17] have shown that the inhalation anesthetics halothane, enflurane, and isoflurane affect both the pharmacodynamics and pharmacokinetics of the calcium channel blocker verapamil in dogs. These anesthetics each affect the hemodynamic response to verapamil to a different degree as indicated by measurements of HR, MAP, cardiac output, left atrial pressure, LV-dP/dt_{MAX}, P-R interval, and systemic vascular resistance.[18] The effects of verapamil on cardiac function are more pronounced in halothane-, enflurane-, and isoflurane-anesthetized dogs than in conscious dogs.[19,20]

Physical Restraint

Another technique often used to obtain measurements is to physically restrain the animal. Physical restraint is employed in both acute studies lasting only a few hours and in long-term studies that require intermittent monitoring over the course of years. Ethical considerations limit any single continuous monitoring session to a few hours. Because restraint itself affects the physiological functioning of the animal, measurement error and variability are introduced into the data. Although diurnal variations can be studied using this technique, it is very labor intensive to do so and therefore expensive. Some species, dogs for example, can be trained to remain in a sling while measurements are taken. With intensive training, dogs can be taught to stand or lie quietly while measurements are taken.[62] However, the training is expensive and perturbations in the data from human interactions may still occur. One of the best-documented problems associated with the use of physical restraint is immobilization stress as can be assessed by evaluation of catecholamines and/or cardiovascular parameters. Stress can be defined as the physiological consequences of some threatening or novel event. Not only immediately stressful events, but the memory of an event can result in a behavioral response. Even though we can predict in general the reaction of the population, the reaction of an individual is never known for sure.[46] Thus, it follows that minimizing or eliminating extraneous stress factors from the experimental design is crucial to a true understanding of the compound, device, or technique being evaluated for its effects. Further, it follows that elimination of sources of variability (stress, for example) may allow fewer animals to be employed with equally valid results.

The effects of immobilization stress resulting from placing an animal in a restrainer are well documented in the scientific literature. For instance, immobilization stress has been shown to markedly affect plasma catecholamine levels. Kvetnansky et al.[29] have found that restraint of Sprague-Dawley rats resulted in increases in norepinephrine (NE) and epinephrine (EPI) of 6 and 40 times, respectively, the values found in undisturbed controls. Peak plasma levels were reached 20 min after application of restraint. The authors also found that even gentle handling of rats for 30 s resulted in a sevenfold increase in NE and a twofold increase in EPI. Transfer of rats from the animal room to the laboratory resulted in a 40% increase in NE and a fourfold increase in EPI.

Livezey et al.[25] have found that plasma EPI and NE in rats peaked 5 min after being restrained. Markedly different responses to stress were found between individuals and between sexes. EPI peaked at 59 and 34 times the baseline value for females and males, respectively. NE peaked at 11 and 14 times the baseline value for females and males, respectively. The authors also found that corticosterone plasma levels climbed to approximately 3 and 1.8 times the baseline level 30 min after application of restraint for females and males, respectively.

Figures 1 and 2 summarize the findings of Kvetnansky et al.,[29] Livezey et al.,[25] Chiueh and Kopin,[32] and Popper et al.[60] on the effects of restraint on plasma catecholamine levels. Blood samples were obtained using indwelling catheters, a technique that has been demonstrated by Kvetnansky et al.[29] to be preferable to decapitation.

Immobilization stress has also been shown to affect certain receptor sites, which in turn can affect responses to pharmacological agents. Restraining rats daily for 2.5 h for seven consecutive days has been shown to desensitize vascular α and cardiac $\beta 1$ receptors.[21] Torda et al.[22] have found that repeated immobilization of rats (2.5 h daily for 40 d) reduced the number of β receptors in the heart, hypothalamus, and brainstem, but not in the spleen. This same protocol of restraint resulted in a decrease in the number of α-1 receptors and an increase in the number of α-2 receptors in the spleen and brainstem, as compared to a control group of unrestrained rats.

Lamprecht et al.[23] found that the BP of normotensive Sprague-Dawley rats was increased by 50 mmHg following 4 weeks of daily immobilization for 2 h. BP did not return to pre-immobilization levels until 2 weeks after cessation of the stressor. The authors also found that serum dopamine-beta-hydroxylase (DBH), an indicator of peripheral sympathetic nerve activity, increased by a factor of two following 4 weeks of repeated immobilization. DBH returned to normal levels 5 d after cessation of the stressor.

Other effects of immobilization stress have also been documented. Sharma[26] found that 8 h of restraint in rats resulted in impaired blood-brain

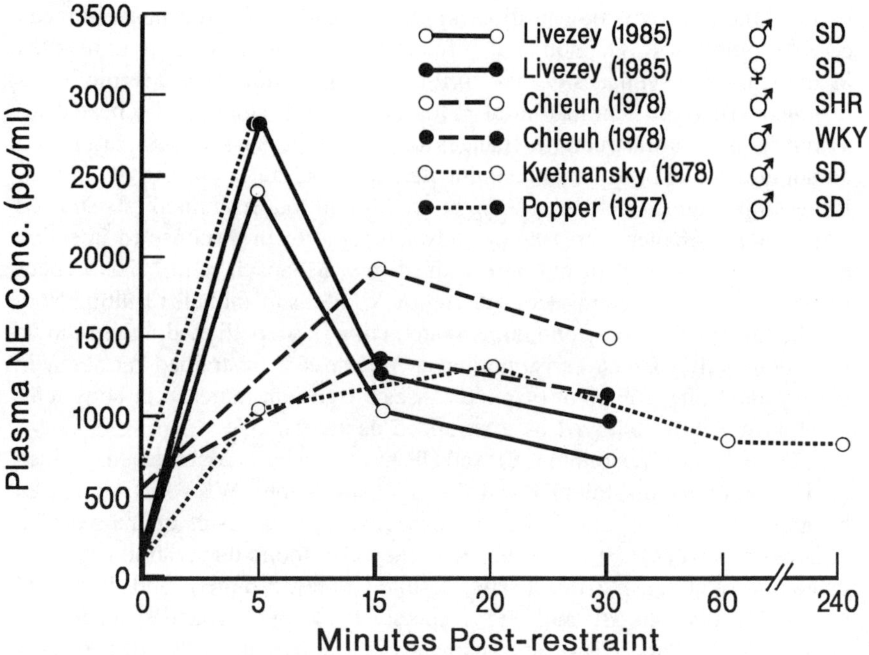

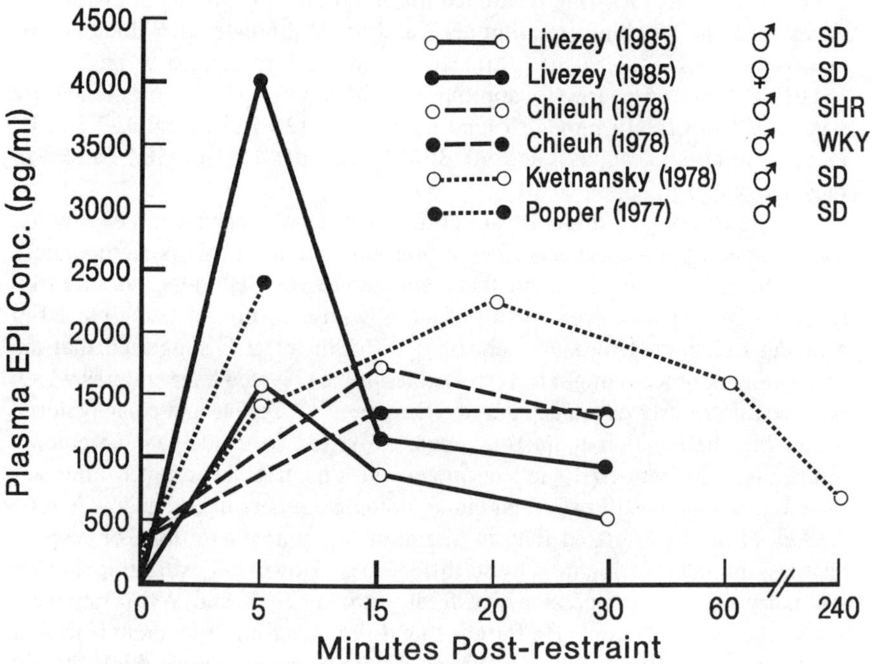

Figures 1 and 2. Results from four studies in rats showing increases in plasma norepinephrine (figure 1) and epinephrine (figure 2) following immobilization.

barrier function, reduction in EEG amplitude, and a 17% reduction in mean cerebral blood flow. Gagnon et al.[27] found that immobilization stress resulted in an increase in synthesis of myofibrillar proteins and increased serum levels of glucocorticoids. Shiraishi et al.[28] found that both single and repeated restraint induced ultrastructural changes in the arterial smooth muscle of rats. Nunomura et al.[24] found that restraint attenuates baroreflex sensitivity in rats. Using implantable telemetry, Long et al.[30] found that restrained rats showed significantly smaller increases in body temperature in response to injection of certain pyrogens than did unrestrained rats. Manoach et al.[36] have documented transient first and second degree A-V blocks in animals resulting from combinations of restraint and human interaction. Tadepalli et al.[2] have shown that total body oxygen consumption is 32% less in restrained rats than in unrestrained rats. Restraint of mice causes a transient increase in body temperature which is followed by a sustained decrease.[3]

The effects of restraint on BP and HR have also been documented. Chiueh et al.[32] found that diastolic BP and HR of Wistar Kyoto (WKY) rats increased by about 10 mmHg and 50 BPM, respectively, as a result of the restraint required for tail cuff BP measurement. They also found that restraining spontaneously hypertensive rats (SHR) resulted in a 25-mmHg and 124-BPM increase in diastolic BP and HR, respectively. Lappe[58] found sustained increases in MAP and HR of approximately 25 mmHg and 150 BPM, respectively, when SHR rats were restrained for 30 min regardless of prior exposure to restraint and handling (see Figures 3 and 4). Malinow et al.[53] documented the average HR of 26 chair-restrained cynomolgus macaques to be 238 $\pm$ 18 BPM. This compares to approximate 24-h average HR monitored by telemetry from freely moving caged animals of 120 BPM[5] and 127 BPM[67] in cynomolgus macaques,[5] and 90 BPM,[64] 120 BPM,[68] and 104 BPM[62] in rhesus macaques.

An interesting example of the utility of this new technology is how the use of telemetry has corrected previous misconceptions of the core temperature of the SHR rat. Bunag[4] found that temperatures of SHR rats measured by rectal probe increased from 37.6 to 38.8°C within 30 min of being placed in a holder of tail-cuff measurement of BP. Collins et al.[47] suggested that the measurement process might be responsible for the elevated temperatures which had been previously documented in the literature.[47-49] Collins and collaborators[47] found that habituation to restraint over 65 d led to a decrease in colonic temperature in both SHR and normotensive rats. However, even after adaptation, SHR rats still exhibited higher colonic temperatures than WKY rats. Berkey et al.[44] postulated that an alteration in sympathetic tone or responsiveness might contribute to these differences. However, using implantable telemetry the core temperatures of freely moving SHR and WKY rats were not statistically different. He further found that handling, moderate restraint, and heat stress caused a greater increase in core temperature in SHR than in WKY rats.

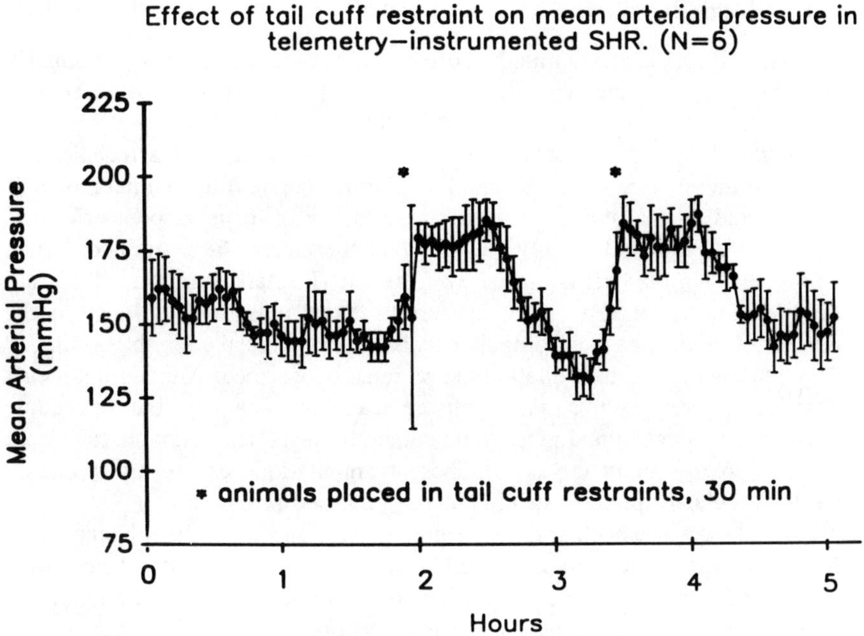

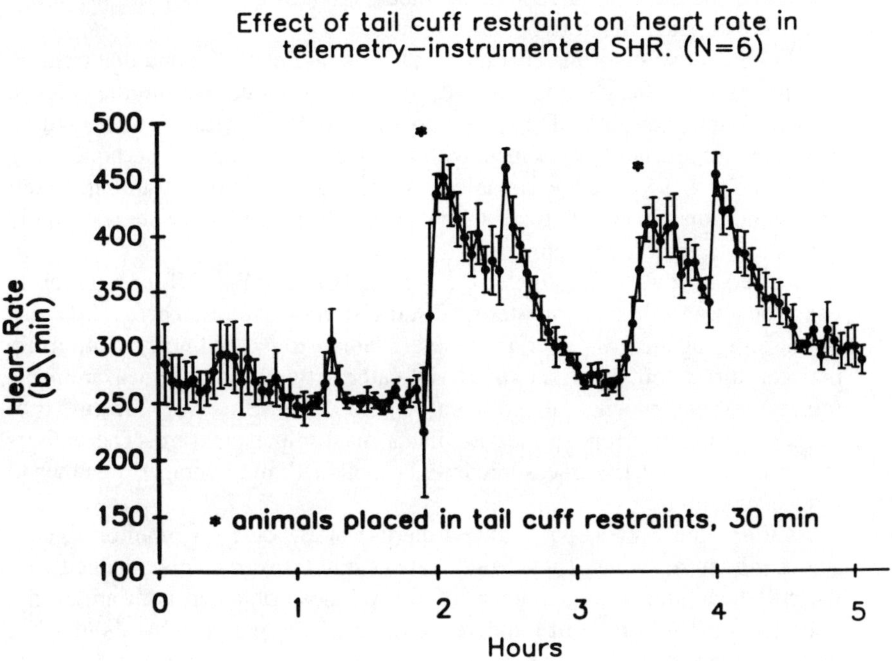

Figures 3 and 4. Effect of restraint on mean arterial pressure (figure 3) and heart rate (figure 4) in telemetry instrumented SHR. Courtesy of Dr. Rodney Lappe, Rorer Pharmaceuticals.

Cuff Blood Pressure

The BP cuff is another method often used to monitor animals, particularly rats. The main advantage of this technique is that it is noninvasive. Several studies have validated the physical accuracy of this method under certain conditions.[4,33,35,37,59] However, the fundamental nature of this technique is such that measurements can be unreliable or inaccurate if the animal moves,[59] is subjected to loud noises or other stressors,[4,33,35,37] or is vasoconstricted.[4]

The cuff method usually requires that the animal be restrained. When applied to rats, the cuff is placed over the tail. The tail artery must therefore be dilated sufficiently to allow sensors to detect pulsations of blood as the cuff is deflated. Rats thermoregulate by modifying caudal artery blood flow.[31] Thus, to assure adequate dilation the rat must be warmed. A modern tail cuff device no longer requires that the rats be heated in a chamber, but for reliable measurements the animal must be restrained for a sufficient time to raise core temperature by about 0.8°C.[4,56] These manipulations can result in altered cardiovascular responses, such as increased blood pressure.[34]

In addition vasoconstrictors, fright, noises, and other disturbances have been shown to reduce or abolish caudal artery pulsations in rats.[4] Under such conditions flow will not be detected at all[4] or it may not be detected until the pressure is significantly below the true systolic pressure.[33] Ferrari et al.[35] found that the tail cuff method often underestimates BP in normotensive rats and overestimates BP in hypertensive rats.

In another study of the cuff technique, Hassler et al.[59] found that training was required for beagle dogs in order to obtain reliable measurements. The authors found that 50% of the dogs could be trained within 1 week and an additional 25% could be trained within 2 weeks. In rhesus macaques, approximately 60% yielded reasonable pressure readings without training. Under ideal conditions, in anesthetized animal preparations, cuff pressure was highly correlated to an arterial catheter ($R = 0.999$).

Petterson et al.[37] investigated the potential use of a BP cuff for monitoring conscious beagle dogs in long-term FDA registration studies. Their validations of the cuff against an arterial catheter demonstrated a linear relationship between direct (obtained via an arterial catheter) and indirect measurements (correlation coefficients ranged from 0.71 to 0.95). However, the cuff produced significant errors in measurements of absolute pressure. The authors recommended that five measurements be obtained and averaged together to reduce variability.

Cimini and Zambraski[38] studied the use of BP cuff for monitoring conscious miniature swine. They found that systolic measurements obtained with the cuff technique were within 4 to 10% of those obtained with an arterial line. The cuff was found to underestimate diastolic pressure by about 17%, while it underestimated MAP by 5%. Correlation between direct and indirect measurements of MAP was 0.72 ($p < 0.01$).

Tethering Systems

Tethering allows continuous monitoring of several different parameters from chronically instrumented conscious animals, including multiple flows, pressures, dimensions, and biopotentials. In addition, this technique also allows for sampling of blood and intravenous infusions using chronic catheters installed as part of the tethering system. The most prominent factor limiting the useful life of this technique is patency of vascular catheters, especially those used for blood withdrawal or pressure measurement. Skilled technicians can maintain patency of catheters in dogs for periods in excess of 1 year, but the useful life in rats is often limited to a few weeks. As with any technique involving exteriorized vascular catheters or lead wires, there is an associated risk of infection.

Tethers have been documented to be a stressor in laboratory animals. Adams et al.[5] have reported that tethered cynomolgus macaques exhibited HR that were 30 to 50 BPM higher than those measured from freely moving monkeys monitored by nonimplanted telemetry. Altered response of the tethered monkeys to propranolol suggested that the increase in HR was due to stimulation of the sympathetic nervous system as a result of tethering. Sylvestri[6] has reported resting HR 100 to 150 BPM higher in tethered rats than in rats monitored by implantable telemetry. In another study, Bohus[63] found 15% differences in HR between tethered and telemetered rats monitored in an open field chamber.

Sheldon et al.[57] found that tethering of telemetry-instrumented SHR rats resulted in an increase in MAP and HR during the first 20 h after attachment to the tether. After acclimatization the two methods provided similar values. Interestingly, despite the similarity in baseline (nontreatment) values, the responses of tethered and untethered rats to clonidine and nifedipine were significantly different. MAP had decreased 18 ± 7 mmHg and HR was increased 25 ± 13 BPM in untethered rats 1 h after administration of nifedipine. There was no significant effect in rate or pressure in the tethered rats. MAP was increased 17 ± 2 mmHg and HR was decreased 129 ± 15 BPM in untethered rats 1 h after administration of clonidine. There was no significant change in MAP 1 h after administration of clonidine to tethered rats, while HR had fallen 150 ± 20 BPM. Both tethered and untethered rats exhibited significant decreases in MAP and HR 8 h after administration of clonidine.

Human interaction by itself has been shown to affect cardiovascular measurements. Totoyan and Cherkovich[64] measured the basal HR of rhesus macaques to be 90 BPM. However, in the presence of human beings HR was elevated to 160 BPM. These monkeys were unrestrained, pretrained, and instrumented with nonimplanted telemetry carried in a jacket. Despite the fact that the monkeys appeared calm following long-term habituation, there was still a significant increase in HR in the presence of human beings.

The potential for inaccurate measurements and the variability associated with conventional techniques have traditionally been dealt with by using large group sizes in experimental protocols. By increasing the number of animals used in a study, statistical significance can be improved in spite of large variability in the collected information. Techniques that improve upon the accuracy of measurements and reduce or eliminate sources of variability have the potential for reducing the number of animals required in a given protocol.

TELEMETRIC MONITORING

A telemetry system for monitoring vital signs consists of two parts, a transmitter and a receiver. The transmitter consists of sensors or lead wires connected to electronic amplifiers similar to those which may be used with a physiograph or other recorder. The amplified signal is used to modulate radio frequency waves which radiate from the animal. The receiver detects the radio waves and reconstructs the original amplified signal.

When used to monitor laboratory animals, telemetry transmitters can either be placed in a jacket or surgically implanted. Placing the trasmitter in a jacket is sometimes employed when monitoring large primates or dogs. This allows for a higher-power, longer-range transmitter and the battery can be easily changed. This is an important factor when monitoring free-ranging animals or animals housed in large cages. This approach requires a high level of maintenance since batteries must be changed, the transmitter may be damaged, percutaneous sites need routine prophylaxis to control infection, and catheters (if employed) require regular flushing.

Surgically implantable telemetry transmitters allow chronic studies to be conducted without the risk of infection from percutaneous wires. Several implantable telemetry systems have been described in the literature.[11,12,39-41] Today, telemetric implants for monitoring cardiovascular data are manufactured commercially.[50,51]

Implantable telemetry as a means of monitoring the vital signs of laboratory animals has a number of advantages when compared to conventional methods,[7,12] including:

- Movement artifacts introduced by cables and commutators associated with tethering systems are eliminated.
- Percutaneous infection is eliminated.
- Stress artifact from tethers, handling, restraint, and human contact is eliminated.
- Altered or disturbed emotional state, sleep, or hibernation are minimized or eliminated.
- Measurements can be obtained without human contact. Thus, telemetry improves the safety of laboratory personnel, particularly when dealing

with primates or protocols involving exposure of the animals to toxic agents.

- Once telemetry devices are implanted little or no maintenance is required.
- Telemetry allows for automated continuous around-the-clock monitoring for days, weeks, or months. This feature allows for a thorough evaluation of time-course profiles including evaluation of biological rhythms and the long-term effects of pharmaceuticals.

As researchers have become more aware of telemetric systems, more sophisticated experiments have been performed that would have been difficult, if not impossible, using conventional means. For example, Broten[61] investigated the autonomic modulation of ultradian BP and HR oscillations in telemetered, free-running dogs. Using Fourier analysis, ultradian oscillations were shown to be prominent in the pressure data with a period of approximately 0.76 cycles per hour. These pressure alterations were approximately twice the magnitude of the diurnal pressure cycles (23 vs. 11 mmHg). Numerous pharmacologic interventions suggest that the autonomic nervous system is primarily responsible for the oscillations and that vagal influences partially dampen the fluctuations.

In another unique experiment, Hassler et al.[62] investigated the arrhythmia potential of new prophylactic treatments for organophosphate (OP) intoxication. ECG transmitters were surgically implanted into ten rhesus macaques. The telemetered ECG signals collected for 45 consecutive days were assessed utilizing Holter arrhythmia analysis techniques. Baseline, prophylactic treatment, exposure to OP, and recovery phase data were analyzed. The data revealed a short-term increase in ventricular ectopic activity associated only with the administration of the OP. During recovery from OP administration the incidence of arrhythmic events rapidly returned to normal levels.

This study also revealed the expected diurnal fluctuations in HR. During the 2-week baseline (before any treatment), the colony average minimum HR was 69 BPM. The baseline 24-h average HR for the colony was 104 BPM. An interesting day of week effect was noted. Average HR were significantly lower on weekends, when there was minimum research activity in the facility, and increased during the week. These alterations suggest increased HR in the animals closely associated with work activity patterns in the research facility, with the highest HR observed on Thursday when the work level in the primate facility was at its peak. Entry of technical staff into the colony could be clearly identified in the HR and activity recordings. HR well in excess of 200 BPM were recorded when humans were in close proximity. This study suggests that significant cardiovascular effects of behavioral origin should be anticipated in study design. These effects may not always be obvious, but related to subtle behavior cues such as extraneous noises related to activity not directly associated with the immediate experiment.[62]

In another example of a rather unique study, Azar and Sanchez-Pena[43] used implantable telemetry to compare changes that occur in BP during prepregnancy, pregnancy, and lactation in normotensive and hypertensive rats. Using telemetry they were able to document changes in BP that occur over the course of both pregnancy and lactation and draw a comparison between hypertensive and normotensive rats. They also found that the presence of an intraperitoneally implanted telemetry transmitter affected neither litter size nor body weight of the pups or mother.

In another study, Morimoto et al.[42] used implantable BP telemetry to study the involvement of prostaglandins in psychological stress-induced responses. The authors found that transferring Wistar rats to a new cage of the same size and design resulted in a 15- to 20-mmHg increase in MAP and a 50- to 70-BPM increase in HR, both of which were sustained for at least 60 min. The authors also found that core temperature increased 0.5 to 0.8°C above temperatures measured in a control group. Inhibition of prostaglandin synthesis with indomethacin suppressed increases in body temperature, but had no effect on stress-induced changes in BP or HR. In yet another study, Mattes and Lemmer[69] demonstrated that the ED_{50} value for amlodipine in rats was seven times lower when administered during the active phase (dark), as opposed to the inactive phase (light).

A typical telemetry system[11] is diagrammed in Figure 5. This system consists of four basic elements: (1) an implantable miniature transmitter which continuously senses, processes, and transmits information from within the animal; (2) a receiver located within or near the cage or pen — most receivers have two antennae oriented at right angles to minimize the possibility of dropouts; (3) a multiplexer, typically located in the animal quarters, that combines the signals from a number of receivers together onto a pair of cables that extend for up to 100 m to the recording system; and (4) a computer-based data-acquisition system which collects, displays, and stores the telemetered data. When pressure data are collected an electronic barometer is required in order to correct the telemetered pressure signal for changes in atmospheric pressure.

This system is designed for monitoring singly housed animals in laboratory cages or other laboratory settings. In order to maximize battery life and minimize package size, the transmitters are designed to operate at very low power over a short distance. All transmitters and receivers operate on the same carrier frequency. Channel separation is obtained by spacing the animals apart or by relying on a steel cage or Faraday shield to contain the radio-frequency transmission. In addition, the receiver circuitry is designed to lock onto the strongest signal available and ignore weaker signals, thereby reducing the chances of cross talk from nearby animals. This design eliminates the need for matching of transmitter and receiver frequencies, reduces the manufacturing cost, simplifies the logistics of the research protocol, and allows a small number of spares to be maintained at a reasonable cost.

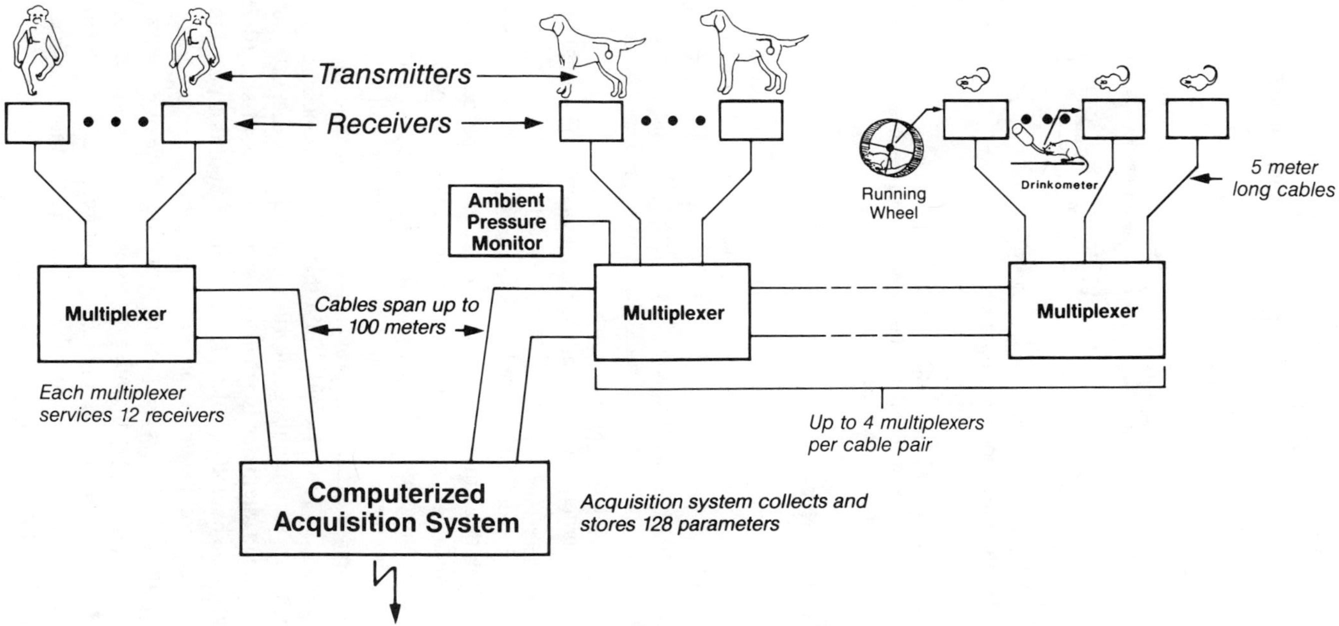

Figure 5. Diagram showing the components of a telemetry system. The ambient pressure monitor is an electronic barometer and is required in order to compensate telemetered pressures for changes in barometric pressure.

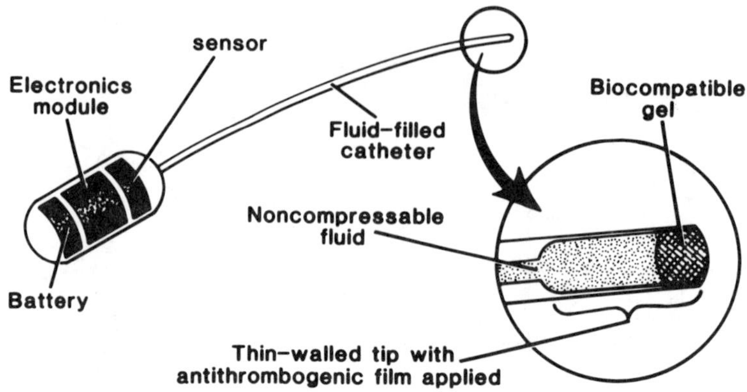

Figure 6. Pressure telemetry implants employ fluid-filled catheters that refer pressure from a specially designed tip to a sensor located in the body of the implant. A gel membrane at the tip prevents blood from entering the lumen of the catheter.

Blood Pressure Transmitter

An implantable pressure transmitter for rats and other small animals is diagrammed in Figure 6. The key components are (1) a sensor; (2) a micropower electronics module to process and digitize the information from the sensor; (3) a battery; and (4) a radio-frequency transmitter. The hermetically sealed housing is constructed of a medical-grade thermoplastic. These devices are precalibrated to three different absolute pressures and are sterilized in ethylene oxide. Built-in magnetic switches allow the devices to be turned on and off either *in vivo* and *ex vivo* to maximize battery service life.

Pressure transmitters have a fluid-filled catheter attached to the sensor located in the body of the transmitter.[11,13] This catheter is composed of two distinct sections. The distal 1 cm consists of a thin-walled thermoplastic membrane, while the remainder of the catheter is composed of a thick-walled, low-compliance urethane. The distal 2 mm of the thin-walled tip is filled with a blood-compatible gel that prevents blood from entering the catheter lumen and retains the low-viscosity fluid that fills the noncompliant section of the catheter. An antithrombogenic film is applied to the exterior of the catheter. The catheter diameter is small compared to vessel size (0.7 mm or 1.2 mm) to minimize hemodynamic effects. Validation of this pressure telemetry system via carotid catheter in rats determined the accuracy to be better than ±5 mmHg in 85% at 3 weeks postimplantation, 86% at 8 weeks, and 78% at 12 weeks.[11] The error in pulse pressure measurement never exceeded 3 mmHg for any transmitter, indicating that relative changes in pressure could still be measured with reasonable accuracy. Of those devices that were not accurate within 5 mmHg, three of them failed during the validation study due to thrombosis, catheter seal leak, or electronics module

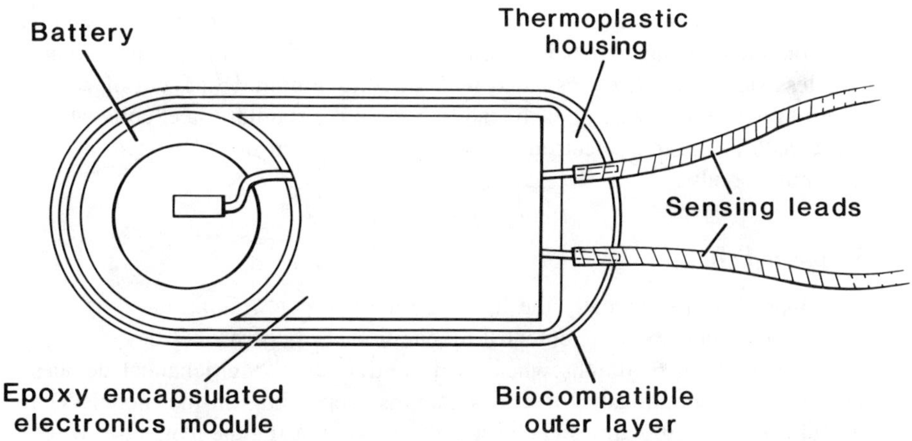

Figure 7. ECG telemetry implants have pairs of flexible stainless steel lead wires extending from the package. These leads can also serve as sensing electrodes.

failure. In another study on anesthetized cynomolgus macaques, Mann et al.[67] found that telemetered pressure measurements were within 5% of those obtained by an arterial cannula 3 to 6 months following implantation.

Another pressure telemetry device for dogs was recently described by Armentano et al.[45] This nonimplantable system employs a Konigsberg (Pasadena, CA) pressure transducer located at the site where pressure is monitored. The lead wires for the transducer were exteriorized and connected to a transmitter carried in a jacket. Locating the transducer at the measurement site eliminates the possibility of errors in measurement due to alterations in fluid-head pressure as the animal changes posture. When using the fluid-filled catheter system described in Reference 11, the laws of physics dictate that a difference in height between the implant body and the catheter tip will result in a measurement error equal to the equivalent head pressure. When the device described in Reference 11 is implanted in a dog as described later in this chapter, the error is negligible when standing, sitting, or lying in a prone position. When implanted in the rat as described in Reference 11 the maximum error due to changes in posture is about 3 mmHg. The location of the implant body relative to the catheter tip should be taken into consideration in order to avoid significant measurement errors. The closer the transmitter body can be located to the tip of the catheter, the smaller the potential fluid-head error.

ECG/Temperature Transmitter

An ECG and temperature implant is shown in Figure 7. Like the pressure device, the ECG implants consist of a hermetically sealed plastic housing that contains the amplifier, battery, and radio-frequency electronics. Extending from the housing are a pair of 0.6-mm-diameter, flexible, helically wound stainless steel lead wires insulated with silicone tubing, 1 or 2 cm of which is removed from the end of each lead to expose the stainless steel coils which serve as the sensing electrodes. A temperature sensor is located in the transmitter body.

Implant Life

Battery life is typically the limiting factor in determining the longevity of a totally implantable telementry preparation. Single-channel devices typically operate for 6 months when used continuously. Multichannel devices will operate continuously for 2 to 4 months, depending on the model. The useful life of the implant can be vastly extended by using the magnetic switch to turn it off when not monitoring the animal or when in storage. In addition, the devices can be recovered at the end of a study and returned to the manufacturer for refurbishment, which typically includes replacement of the battery and other critical components, as well as recalibration and quality-assurance testing.

Surgical Implantation

The implantation procedure varies with the application, species, and transmitter package style. Implants that are available in a flat shape can be placed under the skin or in a body cavity (e.g., the peritoneal cavity). Due to design and manufacturing limitations, however, some rodent-monitoring devices are available only in a cylindrical package, making them suitable only for intraperitoneal implantation.

Following surgical implantation, it is important to allow sufficient time for the animal to recover prior to the study. Surgery itself has been shown to create some degree of stress. In rats, circadian rhythms in HR return within 2 to 4 d following implantation, while BP and activity rhythms return within 4 to 7 d.[66] Popper et al.[60] have shown that when a laparotomy is performed to cannulate the abdominal aorta, an increase in plasma catecholamines occurs. Catecholamine levels peaked on the day after surgery, but by 48 h after the procedure they approached baseline levels.

Following are brief descriptions of how some telemetry devices are implanted in typical applications. The examples below outline procedures that have been proven successful. However, other procedures could be used with equal success.

To monitor a systemic pressure and/or the ECG in the dog or primate, the body of the transmitter is placed in a subcutaneous pocket at the flank. For BP the catheter is tunneled subcutaneously from the flank to the groin, where it is inserted into a branch of the femoral artery. Alternately, the catheter can be inserted directly into the femoral artery. However, the first procedure is preferred, since the impact on blood flow is minimized. The latter procedure is typically used only when a sufficiently large branch is not available. If ECG is monitored, one of the electrodes is usually secured to the tissue near the transmitter body at the left flank. The other sensing lead is routed under the skin and is terminated near the right axilla. To minimize muscle artifact the forward lead can be placed on the surface of the cartilage at the anterior end of the sternum.[52]

To monitor BP in a rat, a laparotomy is required as detailed in Reference 11. The distal 1 cm of the catheter is inserted upstream into the abdominal aorta immediately above the bifurcation and is fixed in position by placing one drop of medical-grade tissue adhesive and a 2-mm-square cellulose fiber patch at the vessel entry site. The body of the transmitter is sutured to the abdominal wall at the incision site.

To monitor ECG signals in rats, transmitters can be placed either subcutaneously or intraperitoneally. Further, devices are available for single- or multiple-lead monitoring. When placed subcutaneously above the scapula, the lead wires are directed under the skin posterior to the forelimbs to small incisions located at the desired electrode sites. The distal 1 mm of the exposed stainless steel coil is inserted into a silicone tip cover to reduce tissue irritation. The electrode is sutured to the underlying tissue to prevent migration. When placing the transmitter body intraabdominally, the procedure is similar except the body is inserted into the peritoneal cavity through a midline incision. The body is sutured to the abdominal wall at the incision and the leads exit the peritoneal cavity through puncture wounds located lateral to the incision.

Data Collection and Handling

A key ingredient to the successful use of radiotelemetry is a computer-based system capable of reliably collecting and storing the vast amount of data that can all too easily be generated by telemetry systems capable of continuous recording. An important consideration is the ability to deal with the possibility of noise in the telemetry link. No biotelemetry system currently available is immune to noise resulting from either dropouts or electromagnetic interference. Since noise in the telemetry link has a different characteristic than biological sources of noise (such as muscle noise on an ECG signal), its effect can be reduced significantly with filtering. Acquisition systems that interface directly with the digital telemetry signal are available.[11,51] In addition, digital-to-analog converters are available that produce an analog voltage from the digital telemetered signal. Systems that can collect cardiovascular data in an analog format are available from a number of companies, including Po-ne-mah (Pinckney, MI), Modular Instruments (Philadelphia, PA), and Buxco (Sharon, CT). Some of these systems incorporate analysis routines, while others provide utilities to export the data to other programs for analysis. For studies conducted in accordance with FDA Good Laboratory Practices, the end user will have to validate the acquisition system to verify the accuracy.

Discussion

Telemetry provides a number of benefits over conventional methods for monitoring BP, temperature, ECG, and other biopotentials. These include elimination of the stress caused by handling, restraint, and tethering; the need for anesthesia during measurement; and the ability to automatically collect data around the clock without the interaction of human beings. For BP, the telemetry system described here provides accurate direct measurements without the problems of clotting and percutaneous infection which are historically associated with traditional vascular catheters.

A potential drawback to telemetry systems is that drug administration and blood withdrawal cannot be accomplished without handling the animal. In this particular aspect, swivel and tether systems may offer an advantage, but only if multiple samples are required in a short time frame. If the vascular system of telemetry-instrumented animals is accessed for blood withdrawal, etc., care must be exercised to neither interfere with the telemetry catheter nor discharge clots during flushing. One of the authors (BPB) has seen evidence suggesting that clots discharged from a vascular catheter located upstream can lodge on the telemetry pressure catheter, resulting in thrombus growth that can surround the tip of the telemetry catheter. If i.v. infusions and blood withdrawal are occasionally required in conjunction with continuous BP monitoring, the blood access location should not be upstream of the telemetry catheter. For example, in dogs, a VAP in the contralateral femoral could be used for both blood withdrawal and occasional validations of telemetry measurements. Another potential advantage of tether systems is that more signals can be monitored than with telemetry. For instance, flow and dimension measurements can be obtained using tethers, but are not routinely available from implantable telemetry.

Since several studies[5,6,63] have shown that tethers appear to impart stress on laboratory animals (as evidenced by elevated HR), the tether as a stressor must be considered in the study protocol. Measurement accuracy of the tether system for monitoring pressure is good, since the accuracy of the transducer can be checked as often as necessary. Fluid-head errors comparable or greater than those observed with the pressure telemetry system of Reference 11 are possible as the animal changes posture. When used for monitoring pressure in rats, the useful life of a tether preparation is usually limited by catheter patency to 6 weeks for skilled technicians.[65] In comparison, Lappe[58] has reported that pressures have been measured in rats using telemetry for periods in excess of 1 year.

Although changes in pressure can be measured accurately in chronic studies, long-term absolute accuracy of the pressure telemetry is of concern because of zero drift of the implanted pressure-sensing element. In anesthetized rats, it has been shown that 78% of telemetry implants are accurate within 5 mmHg 12 weeks after implantation.[11] If a high level of absolute accuracy is required over a long-term protocol, periodic validations of mean pressure should be performed using an arterial stick. Pressure validation at a single mean or perhaps two time points should suffice for most protocols, since drift in scale is highly unusual.[11] If excessive drift is found, calibration values entered in the data-acqusition system can then be modified accordingly. Presently it is clear that the drift of the pressure telemetry is appropriate for studies lasting a few months. However, additional studies are needed to validate the accuracy of the pressure telemetry system for longer studies. As previously mentioned, barometric fluctuations are automatically corrected for and thus do not affect the absolute accuracy of the system.

Compared to methods which employ tethering or physical restraint, the quality of data provided with telemetry appears superior. The literature documents significant physiological changes that can bias the true effect of a compound as a result of immobilization stress. Habituation does not eliminate these effects and they may actually be exacerbated by repeated handling or exposure to restraint.[23]

SUMMARY AND CONCLUSIONS

An examination of the literature has provided several compelling arguments for the implementation of implantable radiotelemetry as an acceptable and even perhaps more appropriate alternative to traditional methods of monitoring laboratory animals. Implantable telemetry has reached a state of evolution whereby accurate and reliable measurements of numerous physiological parameters is possible. The ability of telemetry to obtain measurements in a relatively stress-free manner will provide a more realistic picture of alterations in the physiologic status of the research animal.

Implantable telemetry is now a viable and commercially available alternative to conventional methods for use in pharmacology, toxicology, and general biomedical research. Commercial devices can provide BP, multiple channels of ECG, and temperature simultaneously in freely moving animals in their normal laboratory housing environment. Although the initial cost of telemetry is relatively high, once the device is implanted the longevity and maintenance-free nature of the system more than compensates for the initial cost in many protocols.

The degree of automation already available and the ease with which multiple parameters can be collected allows for an almost indefinite timewise evaluation. Researchers have only begun to probe the utility of this research tool. Totally new insights into the effect of treatments, etc., upon animals will most likely arise. (An analogy can be made to the tremendous improvements afforded in the evaluation of human cardiac patients by the use of long-term Holter analysis.)

The immediate future will provide more sophisticated systems allowing such features as dualchannel pressure monitoring and automated arrhythmia analysis. Further, improved stability of pressure measurements, a wider variety of signals, and longer battery life should be anticipated. Advances in analysis software should be forthcoming to assist the researcher in handling the huge volumes of data which these systems can generate.

REFERENCES

1. Kvetnansky, R., Sun, C. L., Torda, T., and Kopin, I. J., Plasma epinephrine and norepinephrine levels in stressed rats — effect of adrenalectomy, *Pharmacologist,* 19, 241, 1977.
2. Tadepalli, A. S., Walsh, G. M., and Tobia, A. J., Normal cardiac output in the conscious young spontaneously hypertensive rat: evidence for higher oxygen utilization, *Life Sci.,* 15, 1103, 1974.
3. Clement, J. G., Mills, P. M., and Brockway, B. P., Use of telemetry to record body temperature and activity in mice, *J. Pharm. Meth.,* 21, 129, 1989.
4. Bunag, R. D., Facts and fallacies about measuring blood pressure in rats, *Clin. Exp. Hyper. Theory Practice,* A5, No. 10, 1659, 1983.
5. Adams, M. R., Kaplan, J. R., Manuck, S. B., Uberseder, B., and Larkin, K. T., Persistent sympathetic nervous system arousal associated with tethering in cynomolgus macaques, *Lab. Anim. Sci.,* 38, No. 3, 279, 1983.
6. Sylvestri, M., Hoechst Roussel Pharmaceuticals, Dallas, TX. Personal communication.
7. Proceeding of a workshop on blood pressure measurement in hypertensive animal models. National Institutes of Health, Bethesda, MD, DHEW Publ. No. (NIH), September 27, 1977.
8. Livezey, G. T. and Sparber, S. B., Hyperthermia sensitizes rats to cocaine's proconvulsive effects and unmasks EEG evidence of kindling after chronic cocaine, *Pharm. Biochem. Behav.,* 37, No. 4, 761, 1990.
9. Franklin, D., Watson, N. W., and Van Citters, R. L., Blood velocity telemetered from untethered animals, *Nature,* 203, 528, 1964.
10. Deboo, G. J., and Fryer, T. B., Miniature biopotential telemetry system, *Am. J. Med. Elec.,* 4(3), 138, 1965.
11. Brockway, B. P., Mills, P. A., and Azar, S. H., A new method for continuous chronic measurement and recording of blood pressure, heart rate, and activity in the rat via radio-telemetry, *Clin. Exp. Hyper. Theory Practice,* A13, No. 5, 885, 1991.
12. Meindl, J. D. and Ford, A. J., Implantable telemetry in biomedical research, *IEEE Trans. Biomed. Eng.,* BME-31, No. 12, 1984.

13. U.S. Patent 4,846,191. International patents pending.
14. Smith, T. L. and Hutchins, P. M., Anesthetic effects on hemodynamics of spontaneously hypertensive and Wistar-Kyoto rats, *Am. J. Physiol.,* 238 (Heart Circ. Physiol. 7), H539, 1980.
15. Sapru, H. N. and Krieger, A. J., Procedure for the decerebration of the rat, *Brain Res. Bull.,* 3, 675, 1978.
16. Seyde, W. C., McGowan, L., Lund, N., Duling, B., and Longnecker, D. E., Effects of anesthetics on regional hemodynamics in normovolemic and hemorrhaged rats, *Am. J. Physiol.,* 249 (Heart Circ. Physiol. 18), H164, 1985.
17. Chelly, J. E., Hysing, E. S., Abernethy, D. R., Doursout, M. F., and Merin, R. G., Effects of inhalatinal anesthetics on verapamil pharmacokinetics in dogs, *Anesthesiology,* 65, 266, 1986.
18. Chelly, J. E., Hysing, E. S., Hill, D. C., Abernethy, D. R., Dlewati, A., Doursout, M. F., and Merin, R. G., Cardiovascular effects of and interaction between calcium blocking drugs and anesthetics in chronically instrumented dogs. V. Role of pharmacokinetics and the autonomic nervous system in the interactions between verapamil and inhalational anesthetics, *Anesthesiology,* 67, No. 3, 320, 1987.
19. Chelly, J. E., Rogers, K., Hysing, E. S., Taylor, A., Harley, C., and Merin, R. G., Cardiovascular effects of and interaction between calcium blocking drugs and anesthetics in chronically instrumented dogs. I. Verapamil and halothane, *Anesthesiology,* 64, 560, 1986.
20. Rogers, K., Hysing, E. S., Merin, R. G., Taylor, A., Hartley, C., and Chelly, J. E., Cardiovascular effects of and interaction between calcium blocking drugs and anesthetics in chromically instrumented dogs. II. Verapamil, enflurance and isoflurane, *Anesthesiology,* 64, 568, 1986.
21. Yamaguchi, I., Torda, T., Hirata, F., and Kopin, I. J., Adrenoceptor desensitization after immobilization stress or repeated injection of isoproterenol, *Am. J. Physiol.,* 240 (Heart Circ. Physiol. 9), H691, 1981.
22. Torda, T., Kvetnansky, R., and Petrikova, M., Effect of repeated immobilization stress on central and peripheral adrenoceptors in rats, *Endocrinol. Exp.,* 19, 1985.
23. Lamprecht, F., Williams, R. B., and Kopin, I. J., Serum dopamine-beta-hydroxylase during development of immobilization-induced hypertension, *Endocrinology,* 92, No. 3, 953, 1973.
24. Nunomura, T., Nagasak, T., and Shibata, H., Effects of restraint on baroreflex sensitivity in altering heart rate and heat production in rats, *Jpn. J. Physiol.,* 33, 667, 1983.
25. Livezey, G. T., Miller, J. M., and Vogel, W. H., Plasma norepinephrine, epinephrine, and corticosterone stress responses to restraint in individual male and female rats, and their correlations, *Neurosci. Lett.,* 62, 51, 1985.
26. Sharma, H. S. and Dey, P. K., EEG changes following increased blood-brain barrier permeability under long-term immobilization stress in young rats, *Neurosci. Res.,* 5, 224, 1988.
27. Gagnon, J., Ho-Kim, M. A., Boissonneault, G., Dube, J. Y., Rogers, P. A., and Tremblay, R. R., The influence of immobilization stress on cardiac protein synthesis. A possible regulatory role for glucocorticoids, *J. Receptor Res.,* 7, No. 5, 639, 1987.

28. Shiraishi, M., Matsuo, K., and Takebayashi, S., Sequential ultrastructural changes of arteries in rats induced by restrained stress, *Acta. Pathol. Jpn.*, 33, No. 2, 265, 1983.
29. Kvetnansky, R., Sun, C. L., Lake, C. R., Thoa, N., Torda, T., and Kopin, I. J., Effect of handling and forced immobilization on rat plasma levels of epinephrine, norepinephrine, and dopamine-β-hydroxylase, *Endocrinology,* 103, No. 5, 1978.
30. Long, N. C., Morimoto, A., Nakamori, T., and Murakami, N., The effect of physical restraint on IL-1β- and LPS-induced fever, *Phys. Behav.*, 50, 625, 1991.
31. Rand, R. P., Burton, A. C., and Ing, T., The tail of the rat, in temperature regulation and acclimatization, *Can. J. Phys.*, 43, 257, 1965.
32. Chiueh, C. C. and Kopin, I. J., Hyperresponsivity of spontaneously hypertensive rat to indirect measurement of blood pressure, *Am. J. Physiol.*, 234, No. 6, H690, 1978.
33. Borg, E. and Viberg, A., Role of heating in non-invasive blood pressure measurements in rats, *Acta Physiol. Scand.*, 108, 73, 1980.
34. Proskauer, G. G., Neumann, C., and Graef, I., The measurement of the blood pressure in rats with special reference to the effect of changes in temperature, *Am. J. Physiol.*, 143, 290, 1945.
35. Ferrari, A. U., Daffonchio, A., Albergati, F., Bertoli, P., and Mancia, G., Intraarterial pressure alterations during tail-cuff blood pressure measurements in normotensive and hypertensive rats, *J. Hyp.*, 8, No. 10, 1990.
36. Manoach, M., Rubinstein, H., Kauli, N., Karive, N., and Gitter, S., Disturbances in atrioventricular conductivity during stress, *Isr. J. Med. Sci.*, 11, No. 8, 847, 1975.
37. Pettersen, J. C., Linartz, R. R., Hamlin, R. L., and Stoll, R. E., Noninvasive measurement of systemic arterial blood pressure in the conscious beagle dog, *Fundam. Appl. Toxicol.*, 10, 89, 1988.
38. Cimini, C. M. and Zambraski, E. J., Non-invasive blood pressure measurement in Yucatan miniature swine using tail cuff sphygmomanometry, *Lab. Anim. Sci.*, 35, No. 4, 1985.
39. Anderson, R., Livnat, A., and Zehr, J., Crystal-controlled telemetry system for long-term acquisition of physiological data, *Am. J. Phys.*, 244, H162, 1983.
40. Sandler, H., Fryer, T., and Datnow, B., Single-channel pressure telemetry unit, *J. Appl. Physiol.*, 26, No. 2, 235, 1969.
41. Patrick, T., Telemetry of left ventricular diameter and pressure measurements from unrestrained animals, *J. Appl. Physiol.*, 37, No. 2, 276, 1974.
42. Morimoto, A., Watanabe, T., Morimoto, K., Nakamori, T., and Murakami, N., Possible involvement of prostaglandins in psychological stress-induced responses to rats, *J. Physiol.*, 443, 421, 1991.
43. Azar, S. H. and Sanchez-Pena, S., Continuous blood pressure monitoring by radio-telemetry during pregnancy and lactation in spontaneously hypertensive and normotensive rats, in *Hypertension in Pregnancy,* E. V. Cosmi and G. C. Di Renzo, Eds., Monduzzi Editore, SpA, Bologna, Italy, 1991.
44. Berkey, D. L., Meeuwsen, K. W., and Barney, C. C., Measurements of core temperature in spontaneously hypertensive rats by radiotelemetry, *Am. J. Physiol.*, 258 (Regulatory Integrative Comp. Physiol. 27), R743, 1990.

45. Armentano, R., Cabrera-Fischer, E., Breitbart, G., Pichel, R., Levenson, J., and Chau, N. P., Telemetry of aortic pressure in unrestrained animals: validation of the method over a wide range of blood pressure (from 40 to 200 mmHg), *Med. Progr. Technol.*, 16, 125, 1990.
46. Herd, J. A., Cardiovascular response to stress, *Physiol. Rev.*, 71, No. 1, 1991.
47. Collins, M. G., Hunter, W. S., and Blatteis, C. M., Factors producing elevated core temperature in spontaneously hypertensive rats, *J. Appl. Physiol.*, 63, 740, 1987.
48. Isobe, Y., Ohara, K., Kosaka, M., and Aoki, K., Relationships between hypothalamic catecholamines, blood pressure and body temperature in spontaneously hypertensive rats, *Jpn. J. Physiol.*, 30, 805, 1980.
49. Tanaka, C. S., Takaori, S., and Okomoto, K., Central hypothermic effect of L-DOPA in spontaneously hypertensive rats, in *Spontaneous Hypertension: Its Pathogenesis and Complications*, K. Okomoto, Ed., Springer-Verlag, New York, 1972, 83.
50. Konigsburg Instruments, Pasadena, CA.
51. Data Sciences International, St. Paul, MN.
52. Geddes, L. A., Partridge, M., and Hoff, H. E., An EKG lead for exercising subjects, *J. Appl. Phys.*, 15, No. 2, 1960.
53. Malinow, M. R., Hill, J. D., and Ochsner, A. J., III, Heart rate in caged macaca fascicularis. Effects of short-term physical exercise, *J. Med. Primatol.*, 6, 69, 1977.
54. Walsh, G. M., Ferrone, R. A., and Frohlich, E. D., Vasodilation and increased cardiac output during ether anesthesia in the rat, *Pharmacologist*, 19, 194, 1977.
55. McGill, H. C., Functional requirements for blood pressure measurement by research objective, in *Proceedings of a Workshop on Blood Pressure Measurement in Hypertensive Animal Models*, National Institutes of Health, Bethesda, MD, DHEW Publ. No (NIH), September 27, 1977, p. 81.
56. Fregly, M. J., Factors affecting indirect determination of systolic blood pressure in rats, *J. Lab. Clin. Med.*, 62, 223, 1963.
57. Sheldon, J. H., Lappe, R. W., Perrone, M. H., and Cox, B. F., Effects of tethering on the cardiovascular responses to antihypertensive agents in the conscious telemetry-instrumented spontaneously hypertensive rat (SHR), *FASEB J.*, 5. No. 14, 1991.
58. Lappe, R. W., Rorer Pharmaceuticals, Philadelphia, PA. Personal communications, April 1990.
59. Hassler, C. R., Lutz, G. A., Linebaugh, R., and Cummings, K. D., Identification and evaluation of noninvasive blood pressure measuring techniques, *Tox. Appl. Pharmacol.*, 47, 193, 1979.
60. Popper, C. W., Chiueh, C. C., and Kopin, I. J., Plasma catecholamine concentrations in unanesthetized rats during sleep, wakefulness, immobilization and after decapitation, *J. Pharm. Exp. Ther.*, 202, No. 1, 144, 1977.
61. Broten, T. P. and Zehr, J. E., Autoimmune modulation of ultradian blood pressure and heart rate oscillations in dogs, *Am. J. Physiol.*, 256, R1137, 1989.
62. Hassler, C. R., Moutvic, R. R., Hobson, D. W., Lordo, R. A., Vinci, L. T., Dill, G. S., Joiner, R. L., and Hamlin, R. L., Long-term arrhythmia analysis of primates preteated with pyridostigmine, challenged with soman, and treated with atropine and 2-PAM, in Proceedings of the 1989 Medical Defense Bioscience Review, Johns Hopkins University, Columbia, MD.

63. Bohus, B., Telemetered heart rate responses of the rat during free and learned behavior, *Biotelemetry,* 1, 193, 1974.
64. Tatoyan, S. K. and Cherkovich, G. M., Heart rate in monkeys (baboons and macaques) in different physiological states recorded by radiotelemetry, *Folia Primat.,* 17, 255, 1972.
65. Bunag, R. D., Measuring blood pressure in laboratory animals, in *Handbook of Hypertension,* Vol. 14, Blood Pressure Measurement, E. O'Brien and K. O'Malley, Eds., Elsevier Science, Amsterdam, 1991.
66. Brockway, B. P., Data Sciences International, unpublished data.
67. Mann, W. A., Welzel, G., and Kinter, L. B., Determination of resting blood pressure in unrestrained cynomolgus monkeys using implanted telemetric transmitters, *Toxicologist,* 11, No. 1, 335, 1991.
68. Carlson, K. H., Wolff, R. K., Tielking, R. L., Franklin, P. A., Yeager, H. K., and Dorato, M. A., Real-time measurements of cardiac effects during inhalation exposures of monkeys or dogs, *Toxicologist,* 11, No. 1, 1991.
69. Mattes, A. and Lemmer, B., Effects of amlodipine on circadian rhythms in blood pressure, heart rate, and motility: a telemetric study in rats, *Chron. Intl.,* 8, No. 6, 526, 1991.

Chapter

10

Statistical Considerations in Approaches for the Reduction and Replacement of Animal Usage for Toxicity Testing

Shayne C. Gad
Director
Medical Affairs Technical Services
Becton Dickinson
Research Triangle Park, North Carolina

Statistics serve an essential yet frequently misunderstood role in toxicology. This essentially is in meeting the need to have a known level of confidence that something is different from (or the same as) something else. This confidence is traditionally achieved by using one of a set of techniques for hypothesis testing (such as t-tests, analysis of variance, and chi square tests).

Just as essential, but less well understood, is the use of statistics for understanding what data means — the modeling function. An aspect of this is the use of statistics to design better and more efficient tests. It is particularly in these last two uses, and in using some less traditional statistical usage in toxicology testing, that are of interest in this chapter.

This chapter will not attempt to comprehensively address the use of statistics in toxicology — this is already the subject of several tests (Gad and Weil, 1988; Krewski and Franklin, 1991; and Salsburg, 1986). Rather, here we will review some fundamental concepts and look at statistical ways and methods for reducing animal usage and refinement of uses.

0-8493-8896-1/93/$0.00 + $.50

FUNDAMENTALS

A major difficulty in toxicology and in the development and utilization of new test systems is that few of the scientists involved are well versed in statistics, as are few of the statisticians who are called upon to help grasp the biological realities, the nature of the questions being asked, and the implications of a problem before designs and methods of analysis are selected and employed.

As a point of departure, it is essential that any analysis of study results be interpreted by a professional who firmly understands both the difference between biological significance and statistical significance and the nature of different types of data. For the first of these, we should consider the four possible combinations of these two types of significance, revealing the relationships below.

		Statistical significance	
		No	Yes
Biological significance	No	Case I	Case III
	Yes	Case II	Case IV

Cases I and IV give us no problems, for the answers are the same statistically and biologically. But cases II and III present problems. In case II, we have a circumstance where there is a statistical significance in the measured difference between treated and control groups, but there is no true biological significance to the finding. This is not an uncommon happening, for example, in the case of many *in vitro* test system parameters. In case III, we have no statistical significance, but the differences between groups are biologically/toxicologically significant. An example of this second situation is when we see a few cases of a very rare observation in a test system. In both of these latter instances, numerical analysis, no matter how well done, is no substitute for professional judgment. Here, perhaps, is the cutting edge of what really makes a practicing toxicologist. Along with this, however, must come a feeling for the different types of data and of the implications involved with them.

In all areas of biological research, optimal design and appropriate interpretation of experiments require that the researcher understand both the biological and technological underpinnings of the system being studied and of the data being generated. From the point of view of the statistician, it is vitally important that the experimenter both know and be able to communicate the nature of the data, as defined below in Table 1 (or in any number of similar schemes).

The nature of the data collected is determined by three considerations. These are the biological source of the data (the system being studied), the

TABLE 1
Types of Data and Examples of Each Type

Classification	Type	Example[a]
By scale		
Continuous	Scalar	Neutral red measurements
	Ranked	Draize eye scores
Discontinuous	Scalar	Weeks until the first observation of a tumor in a carcinogenicity study
	Ranked	USP-style cytotoxicity classifications (0–4)
	Attribute	Cell types
	Quantal	Dead/alive or present/absent
By distribution	Normal	Body weights
	Bimodal	Some clinical chemistry and biochemical parameters
	Others	Measures of time cell transformation

[a] It should be kept in mind that though these examples are most commonly of the data types assigned above, it is not always the case.

instrumentation and techniques being used to make measurements, and the design of the experiment. The researcher has some degree of control over each of these — the least over the biological system (he/she normally has a choice of one of the several models to study) and the most over the design of the experiment or study. Such choices, in fact, dictate the type of data generated by a study.

Statistical methods are each based on specific assumptions. Parametric statistics, which are most familiar to the majority of scientists, have more stringent sets of underlying assumptions (for many parametric statistical methods, such as analysis of variance, for example, it is that the data are continuous). This means that, at least theoretically, the data can assume any of an infinite number of values between any two fixed points.

Limitations on our ability to measure constrain the extent to which the real-world situation approaches the theoretical here, but many of the variables studied in toxicology are in fact continuous. Examples of these are lengths, weights, concentrations, temperatures, periods of time, and percentages. For these continuous variables we may describe the character of a sample using measures of central tendency and dispersion with which we are most familiar — the mean, denoted by the symbol X and also called the arithmetic average (calculated by adding up all the values in a group then dividing by N), and the standard deviation (SD), which is denoted by the Greek letter sigma (σ).

Contrasted with this continuous data, however, we have discontinuous (or discrete) data, which can only assume certain fixed numerical values. That is, no intermediate values are possible between the ''fixed'' points in discontinuous data. In these cases our selection of types of statistical tools or tests is, as will be explained later, somewhat more limited.

FUNCTIONS OF STATISTICS

Statistical methods may serve to do any combination of three possible tasks. The one we are most familiar with is hypothesis testing — that is, determining if two (or more) groups of data differ from each other at a predetermined level of confidence. A second function is the construction and use of models, which is most commonly seen as linear regression or as the derivation of some form of correlation coefficient. Model fitting allows us to relate one variable (typically a treatment or "independent" variable) with other variables (usually one or more effects or "dependent" variables). The third function, reduction of dimensionality, is still much less commonly utilized than the first two. In this final category are the methods for reducing the number of variables in a system which only minimally reduce the amount of information, therefore making a problem easier to visualize and to understand. Examples of such techniques are factor analysis and cluster analysis. A subset of this last function, discussed under descriptive statistics, is the reduction of raw data to single expressions of central tendency and variability (such as the mean and SD).

There is also a special subset which is part of both the second and third functions of statistics. This is data transformation, which includes such things as the conversion of numbers to log or probit values. There are also some special-case methods which can serve several of these purposes at once. Much of this chapter will look at these functions.

First, however, a short review of experimental design is called for as it is key to developing new (or improving existing) test systems.

EXPERIMENTAL DESIGN

Both the cost to perform research and the value that society places upon the results of such efforts have continued to increase rapidly. Especially as this does not seem likely to become less the case in the foreseeable future, it is essential that every experiment and study yield as much information as possible, and that (more specifically) the results of each study have the greatest possible chance of answering the questions it was conducted to address. *A priori* selection of statistical methodology (as opposed to the *post hoc* approach) is as significant a portion of the process of protocol development and experimental design as any other and can measurably enhance the value of the study. Such prior selection of statistical methodologies is essential for effective detailing of such other portions of a protocol as the number of animals per group and the sampling intervals for body weight. Implied in such a selection is that the toxicologist has both an in-depth knowledge of the area of investigation and an understanding of the general principles of experimental design, for the analysis of any set of data is dictated to a large extent by the manner in which the data are obtained.

The four statistical principles of experimental design are replication, randomization, concurrent (''local'') control, and balance. In short form, these are summarized as follows.

Replication. Any treatment must be applied to more than one experimental unit (animal, plate of cells, etc.). This provides more accuracy in the measurement of a response than can be obtained from a single observation, since underlying experimental errors tend to cancel each other out. It also supplies an estimate of the experimental error derived from the variability among all of the measurements taken (*replicates*).

Randomization. This is practiced to ensure that every treatment will have its fair share of extreme high and extreme low values and allows the toxicologist to proceed as if the assumption of ''independence'' is valid. That is, there is no avoidable/known systematic bias in how one obtains data.

Concurrent control. Comparisons between treatments should be made, to the maximum extent possible, between experimental units from the same group. This is, animals used as a control group should come from the same source and lot as test group animals. And except for the treatment being evaluated, test and control animals should be maintained and handled in exactly the same manner.

Balance. If the effect of several different factors is being evaluated simultaneously, the experiment should be laid out in such a way that the contributions of the different factors can be separately distinguished and estimated.

The goal of all these principles is good statistical efficiency and the economizing of resources (animal and otherwise). An alternative way of looking at this in a step-wise logical manner is to do a logic flow analysis of the problem. Such an undertaking is conducted in three steps and should be performed every time any major study or project is initiated. These steps are detailed below.

1. *Define the objective of the study or experiment — get a clear statement of what questions are being asked.*
   Can the question, in fact, be broken down into a set of subquestions?
   Are we asking one or more of these questions repeatedly?
   For example, dose ''X'' develop at 30, 60, 90 + days and/or does it progress/regress or recover?
   What is our model to be in answering this/these question?
   Is it appropriate and acceptably sensitive?
2. *For each subquestion (i.e., separate major variable to be studied):*
   How is the variable of interest to be measured?
   What is the nature of the data generated by the measure?
   Are we getting an efficient set of data? Or are we buying (a) too

little information, and would another technique improve the quality of the information generated to the point that it becomes a higher "class" of data? or (b) too much information, i.e., does some underlying aspect of the measure limit the class of data obtainable within the bounds of feasibility of effort?

Are there possible interactions between measurements?

Can they be separated/identified?

Is our N (sample size) both sufficient and efficient?

What is the control — formal or informal? Is it appropriate?

Are we needlessly adding confounding variables (asking inadvertent or unwanted questions)?

Are there "lurking variables" present? These are undesired and not readily recognized differences which can affect results, such as different technicians observing different groups of animals.

How large an effect will be considered biologically significant? This is a question that can only be resolved by reference to experience or historical control data.

3. *What are the possible outcomes of the study — i.e., what answers are possible to both our subquestions and to our major question?*

   How do we use these answers?

   Do the possible answers offer a reasonable expectation of achieving the objectives that prompted us to initiate the study?

What new questions may these answers lead us to ask? Can the study be redesigned, before it is actually started, so that these "revealed" questions may be answered in the original study?

A practical example of the application of this approach can be demonstrated in the process of optimizing and validating an *in vitro* eye irritation assay. Although in such a situation the primary question being asked is usually "does the test system predict what happens in man?", even at the beginning there are a number of other questions that it is expected that the study will answer. Two such are (a) if irritation is predicted, what is the relative severity of it? and (b) are there other measures of toxicity which could improve assay performance? Several, if not all, of these three questions are actually to be asked repeatedly during the course of the project. Before the study starts, a plan and arrangements must be formed to make measurements to allow us to answer these questions.

When considering the last portion of our logic analysis, however, we must start by considering each of the things that may go wrong during the study. These include the occurrence of a contamination of cells (do we continue or start over? Will we be able to separate the portions of observed effects that are due to the chemical under study and those that are due to secondary?), finding that extreme variability is occurring in control culture function, or revealing the existence of some hidden variable. Can we preclude the possibility of a disease or yeast contamination by doing a more extensive microbiological screening and quarantine on our source cells prior to the start of the study? Could we select a better test model — one that is not as variable in response and yet is as sensitive as desired?

One last aspect of experimental design should be addressed here, however: determining sufficient test and control group sizes to allow us to have an adequate level of confidence in the results on an analysis. The ways of calculating these are reviewed by Gad and Weil (1988) and by Krewski and Franklin (1991). What should be kept in mind, however, is that prior knowledge of test performance can be used to make the decision tools involved more efficient.

SCREENING

Much of the toxicity (or ''safety'') testing that is performed is meant to function as a screen. This is particularly true for short-term or acute tests for which *in vitro* alternatives are now being most actively pursued. In an early (primary or first-stage) screen, a relatively large number of compounds will be tested. It is unlikely that one will stand out so much as to be statistically significantly more important than all the other compounds. A more or less continuous range of activities will be found. Compounds showing the activity of interest will proceed to the next assay or tier of tests in the series or may pass through undetected and may be used as lead compounds in a new cycle of testing and evaluation. Undesirable activity will lead to rapid dropping of most candidate compounds from commercial development or further consideration.

Each test or assay can have an associated activity criterion. If the result for a particular test compound falls below this criterion, the compound may pass to the next stage of development (i.e., it will pass the screen). This criterion could be based on statistical data (i.e., all compounds with observed activities significantly greater than the control at the 10% probability level could be tagged). However, for early screens, such a criterion may be too strict and few compounds may be so identified as active.

Statistical modeling of the assay system may lead to the improvement of the design of the system to reduce the interval between discoveries. The objectives behind a screen and considerations of (1) costs for producing compounds and testing and of (2) the degree of uncertainty about test performance will determine desired performance characteristics of specific cases. Preliminary results suggest that in the most common case of early toxicity screens performed to remove possible problem comopounds, it may be beneficial to increase the number of compounds tested, decrease the numbers of test units (such as animals) per group, and increase the range and number of doses. The result will be less information on more structures, but an overall increase in the frequency of discovery (assuming that truly toxic compounds are entering the system at a steady rate). The ultimate measure of success, of course, is a reduction of the number of compounds found to be neurotoxic in humans.

It should be noted that the methods described here are better suited to analyzing screening data when the interest is truly in detecting the absence of an effect with little chance of false negatives. Anderson and Hauck (1983) should be consulted for a more thorough discussion of direction of considerations involved in cases where the desire is to have few false positives (i.e., when the direction of the hypothesis test has been reversed). There are also many forms of graphical analysis methods available, including some newer forms that are particularly well suited to multivariate data (the type that is common in more complicated screening designs).

It can be assumed for any one test or toxicity endpoint that a compound is either active or inactive and that the proportion of actives can be estimated from past experience. After testing, a compound will be classified as positive or negative. It is then possible to design the assay so as to optimize the following characteristics:

1. Sensitivity: the ratio of true positives to total actives
2. Specificity: the ratio of true negatives to total inactives
3. Positive accuracy: the ratio of true to observed positives
4. Negative accuracy: the ratio of true to observed negatives
5. Capacity
6. Reproducibility

These interact as follows:

	Agent tested	
Screen outcome	Positive	Negative
Positive	a	b
Negative	c	d

Discovery (sensitivity) = a/(a + c) where a = toxicants are found positive
Discrimination (specificity) = d/(b + d) where d = all nontoxicants are found negative b + d = all nontoxicants tested

The design of each assay and the choice of the activity criterion should therefore be adjusted, bearing in mind the relative costs of retaining false positives and rejecting false negatives. Decreasing the group sizes in the early assays reduces the chance of obtaining significance at any particular level (such as 5%) so that the activity criterion must be relaxed, in a statistical sense, to allow more compounds through. At some stage, however, it becomes too expensive to continue screening many false positives, and the criteria must be tightened up accordingly. Where the criteria are set depends on what acceptable noise levels are in a screening system. As previously pointed out, there are several classes of screens.

The basic factors that influence the statistical performance of any bioassay in terms of its sensitivity and error rates have been reviewed by Gad (1989).

In 1987, Healey presented a review of the factors that influence the power (the ability to detect dose-related effect when such actually exists) of a study. In brief, the power of a study depends on seven aspects of study design:

1. Sample size
2. Background variability (error variance)
3. Size of true effect to be detected (i.e., objective of the study)
4. Type of significance test
5. Significance level
6. Decision rule (how many false positives will one accept?)

There are several ways to increase power, each with consequences:

1. Increase the sample size	Greater resources required
2. Design test to detect larger differences	Less useful conclusions
3. Use a more powerful significance test	Stronger assumptions required
4. Increase the significance level	Higher statistical false positive rate
5. Use one-tailed decision rule	Blind to effects in the opposite direction

Timely and constant incorporation of knowledge of test system characteristics and performance will serve to reduce background variability and allow sharper focus on the actual variable of interest.

Screens may thus truly be considered the biologic equivalent of exploratory data analysis (EDA). The function of both is to discover items of interest for more intensive analysis. EDA methods, in fact, provide a number of useful possibilities for less rigid and yet quite utilitarian approaches to the statistical analysis of the data from screens and form the basis for some of the alternative approaches presented and evaluated in this article.

Only the special cases of *in vitro* or other alternative assays are discussed here. However, the evaluation and approaches should be valid for all similar screening data sets regardless of source. The methods are not sensitive to the biases induced by the degree of interdependence found in many screening batteries that employ multiple measures.

It is, however, important to distinguish that there are both univariate and multivariate cases to be considered for many screens. The generally used methods of analyzing tend to treat all cases as univariate and analyze accordingly. However, many of the test systems used as screens in toxicology are of the multi-end point, or shotgun, nature. They measure multiple variables (which are seemingly independent or at least have no apparent direct dependence or linear correlation) at multiple points in time. In most such cases, although analysis of each of the individual measures does not give a biased appreciation of the data, it does lose a lot of information (and, therefore, power in the sense of both sensitivity and discrimination) that is present in the conjunction of and relationships between the multiple measures.

Two alternative approaches to the statistical analysis of screen data which do not seem to have the limitations and drawbacks of classic methods are the control chart approach and the central tendency plot (a semigraphical exploratory data analysis method). These are not new methods that probably would serve just as well, but the two will serve as examples and have been successfully employed.

Control charts. The control chart approach, commonly used in manufacturing quality control for another form of screening, offers some desirable characteristics. During the development of screen methodology, for example, by keeping records of cumulative results, an initial estimate of the variability, such as SD, of each assay in available when full-scale use of the screen starts. The initial estimates can then be revised as more data are generated (i.e., as we become more familiar with screen performance).

The following example shows the usefulness of control charts for control measurements in a screening procedure. Our example test for screening potential cytotoxic agents measures remaining metabolic function in cells exposed to test compounds compared to a control treatment. A control chart was established to monitor the performance of the control agent (1) to establish the mean and variability of the control and (2) to ensure that the results of the control for a given experiment are within reasonable limits (a validation of the assay procedure). The average activity and the average range for a series of experiments are presented in Table 2.

As in control charts for quality control, the mean and average range of the assay were determined from previous experiments. In this example, the screen had been run 20 times before the data shown. These initial data showed a mean activity of 400 units and a mean range (R) of 90. These values were used for the control chart in Figure 1. The subgroups were of size 5. The action limits for the X and range charts were calculated as follows:

Using the factor (0.58) from column A (Table 2) for a *n* of 5, we get **X** ± 0.58 **R** = 400 ± 0.58 (90) = 348 to 452 (X chart). Then, using the upper limit (Du) for an n of 5 (which is 2.11), R (2.11) = 90 (2.11) = 190, the upper limit for the range.

These are used to establish the control chart shown in Figure 1, which also shows the individual data points. Note that the range limit actually establishes a limit for the variability of our data — that it is, in fact, a detector for the presence of outlier (extreme) values.

TABLE 2
Average Enzyme Activities and Ranges for Screening Procedures

Test no.	Mean	Range	Test no.	Mean	Range
1	380	40	11	280	120
2	430	30	12	410	100
3	340	30	13	400	220
4	480	60	14	340	50
5	380	240	15	370	40
6	450	40	16	430	140
7	490	50	17	370	60
8	320	90	18	450	80
9	480	50	19	320	70
10	340	80	20	420	130

Values for determining upper and lower limits for mean (X) and range charts

		Range chart factors	
Sample size of subgroup (n)	Column A: factor for X chart	Lower limit (D_I)	Upper limit (D_U)
2	1.88	0	3.27
3	1.02	0	2.57
4	0.73	0	2.28
5	0.58	0	2.11
6	0.48	0	2.00
7	0.42	0.08	1.92
8	0.37	0.14	1.86
9	0.34	0.18	1.82
10	0.31	0.22	1.78
20	0.18	0.41	1.59

Such charts may be constructed and used for proportion or count (attribute) type data. By constructing such charts for the range of incidence of one or more unusual observation types from control data, we may then use them as rapid and efficient tools for detecting effects in groups being assessed for that same screen end point.

Central tendency plots. The objective behind our analysis of any screen data is to have a means of efficiently, rapidly, and objectively identifying those agents that have a reasonable probability of being active. Any materials that we so identify may be further investigated in a more rigorous manner that will generate data that can be analyzed by traditional means. In other words, we want a method that makes results that are out of the ordinary stand out. To do this, we must first set the limits on ordinary (summarize the control case data), then overlay a scheme that causes those things that are not ordinary to become readily detected (exposed, in EDA terms). One can then proceed to perform confirmatory tests and statistical analysis, using traditional hypothesis testing techniques, if so desired.

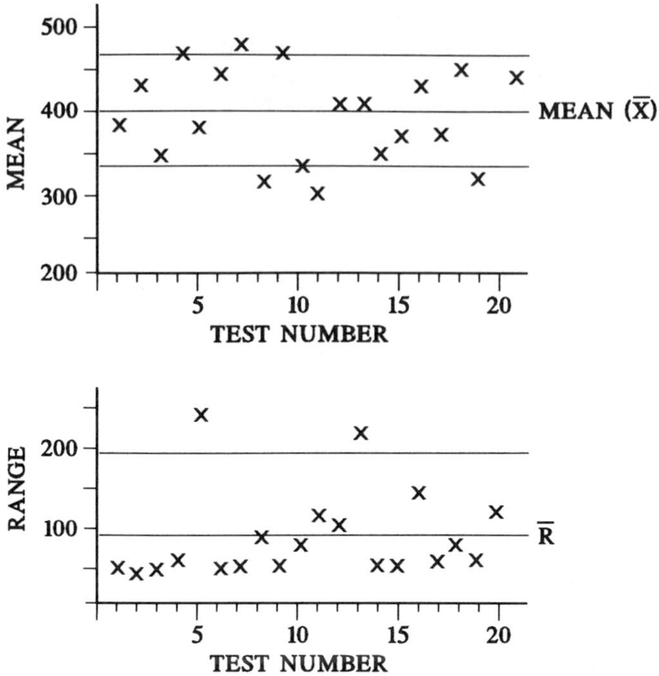

Figure 1. Control chart for means and range for control group in screening procedure for agents affecting cell metabolic activity.

If we proceed to collect a set of control data on a variable (e.g., our observation of cell lethality scores) from some number of control plates and then plot it as a set of two histograms (one for individual plates and the second for the highest total score in each randomly assigned group of five plates), we would have two histograms (Figure 2). Such a plot then acts to identify the nature of our data, visually classifying it into those that will not influence the outcome of an analysis. In so doing, the position of control (normal) observations is readily revealed as a central tendency in the data (hence the name for this technique).

We can and should develop such plots for each of our variables. Simple inspection makes clear that answers having no discriminatory power (0 values in the example central tendency plot) do not interest us or influence our identifying of an outlier in a group and should simply be put aside or ignored before continuing on with analysis. This first stage of summarizing the control data thus serves to give us a device for identifying data with discriminatory power (extreme values) and for then setting aside other data (for purposes of identifying compounds of interest).

Focusing our efforts on the remainder, it becomes clear that although the incidence of a single low nonzero observation in a group means nothing

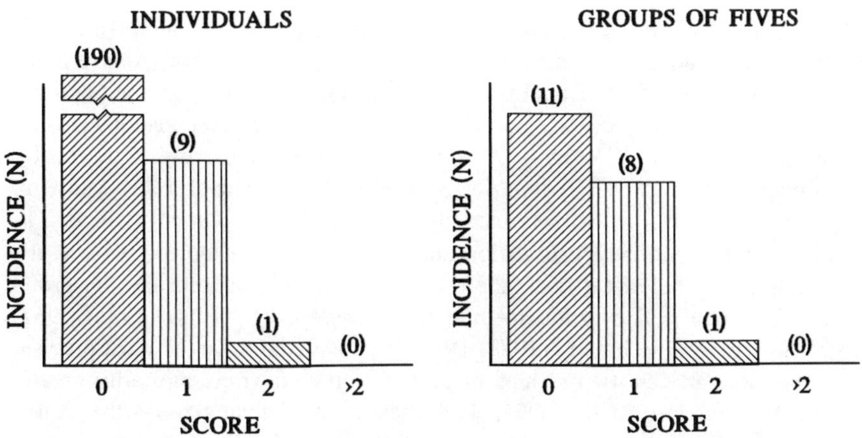

Figure 2. Example of central tendency plots constructed to allow data screening. Data shown are cell-lethality scores from 200 control plates, plotted as individuals and as randomly assigned groups of five. Possible individual scores range from 0 to 4. Group total scores thus range from 0 to 20. Shown to the right are the number of groups that contain individual scores in the indicated categories.

(i.e., it occurs 40% of the time by chance), scores of 2 or more occurred less than 5% of the time by chance. So we can simply perform an extreme value screen on our collapsed data sets, looking for total group values or individual values that are beyond our acceptance criteria.

The approach of this method is then to develop a histogram for each ranked on quantal variable, by both individual and group. Useless data (that which will not influence the outcome of the analysis) are then identified and not used further in analysis. Group scores may then be simply evaluated against the baseline histograms to identify those groups with scores divergent enough from control to either be true positives or an acceptably low incidence false positive. Additional control data can continue to be incorporated in such a system over time, both increasing the power of the analysis and providing a check on screen performance.

LETHALITY TESTING

The precision with which lethality curves are described will depend on the number of groups (hence, dosages) and the number of animals in each group. Between 1940 and 1980, the standard was to use from four to six dosages with ten animals per dosage. Some regulatory guidelines even called for achieving 95% confidence limits within $\pm 10\%$ of the calculated value. The current emphasis is on limiting the number of animals used for lethality testing, particularly with recognition of the limited value of such "precise" data. Hence, the number and size of dosage groups will depend, to an extent, on the methods of statistical analysis. The classic statistical methods for analyzing lethality data (or, indeed, any quantal dosage response data) were published between 1930 and 1960 and have been extensively reviewed by Armitage and Allen (1950) and Morrison et al. (1968). The probit and moving average methods currently remain the most commonly used methods. In general, all methods of calculation and computation are more precise if the designs are symmetrical (i.e., the dosages are evenly spaced and the group sizes are equal). The probit method first developed by Bliss (1935, 1957) and later refined by Finney (1971, 1985) is considered to be the most precise, but requires at least two groups of partial responses (i.e., mortality greater than 0%, but less than 100%). This may require dosing more than three groups until this criteria is met. It also does not deal effectively with groups that had either 0 or 100% lethality (The most common correction for these is to substitute 0.1 for 0% and 99.7 for 100%). The moving average method, first described by Thompson and Weil (1952), does not require partial responses and deals effectively with complete responses and can therefore produce an acceptable estimate of an LD_{50} with as few as three groups of three to five animals each. This method requires that the dosage be separated by a constant geometric factor (e.g., 2, 4, and 8 mg/kg) and that groups be of equal size. Weil (1952), and later Gad and Weil (1988), have published tables that allow for the easy calculation of the LD_{50} with $K = 3$ (four dosage groups). The LD_{50} for $K < 3$ can be easily calculated without the aid of tables. In addition, methods for estimating the confidence limits of this calculated LD_{50} have also been published (Gad and Weil, 1988). Traditionally, the moving average method has not been more extensively used because, while it yielded an estimate of the LD_{50}, and slope can be obtained from as few as three groups and five animals per group, providing at least one group shows a response less than 50% and another shows a response was greater than 50%.

The normit-chi^2 square, developed by Berkson (1955), it also sometimes used. Like the probit method, the normit-chi^2 does not absolutely require equally spaced dosages or equal group sizes, but does require at least one partial response. Hence, using the normit method, chi^2 method, fewer dosage groups may be needed than with the probit method. According to Waud (1972), the correction for including complete responses is better than that used for probit analysis, but is still "tainted". His method supposedly deals adequately with complete responses, but is extremely complex, and probably for this reason is rarely used.

In an early paper, Karber (1931) published a simple method which has been often described, but rarely cited, for calculating an LD_{50}. It does not require that dosages be equally spaced, but widely divergent dosages will lead to a biased result. The method was originally described for groups of equal size, but groups of slightly different size can be used, providing that they do not differ by more than a few animals each. In this case, mean group size can be inserted into Karber's formula with little change in accuracy. The formula is very simple and one can calculate an acceptable estimate of the LD_{50} quickly with only a few arithmetic computations. This method, unlike those mentioned above, does not allow for calculating the confidence limit or slope of the probit response curve. Hence, if these calculated parameters are not sought, this method allows one a bit more freedom in picking dosages.

While much has been written about the influence of gender on acute lethality, most authors now agree that there is seldom any substantial difference in the LD_{50} due to sex (DePass et al., 1984). In those instances where there is a sex-related difference, females tend to be more sensitive than males. If one is willing to accept this amount of uncertainty, only one sex needs to be tested. Alternatively, as few as two to three animals per sex dosage can be used. Schutz and Fuchs (1982) have demonstrated that by pooling sexes there is seldom any substantial difference in the LD_{50} calculations with five per sex vs. three per sex groups. If there are no substantial differences between sexes (i.e., 0% mortality for males and 100% for females at a dosage), the results from the sexes can provide a pooled LD_{50}. For most safety considerations, an LD_{50} derived on this basis will be acceptable and will result in the use of fewer animals.

A wide variety of designs for acute toxicity studies have been developed to address a range of needs and definitions of efficiency (see Gad and Chengelis, 1988, for a complete review). Two of particular note which use very few animals to fulfill defined information needs are the up-and-down procedure for lethality (Bruce, 1985) and the staggered start heavy acute study for characterization of system toxicity (Gad et al., 1984).

OCULAR IRRITATION

The "standard" test here is the rabbit eye irritation study, commonly (and mistakenly) labeled the Draize test. The performance of this test has undergone significant critical review, resulting in marked changes in test methodology for the *in vivo* test. Additionally, a great deal of work has been done (and is ongoing) on the development and "validation" of *in vitro* test systems. The contributions that the statistical approaches and thinking overviewed in this chapter can make to the continued reduction and refinement of animal usage for eye irritation testing include the following:

- By careful consideration of the objectives behind test conduct, the need for testing may be reduced and more attainable goals for *in vitro* system performance may be identified.

- Careful analysis of test performance can lead to reductions in the number of animals employed per test while also allowing refinement of test design to a more humane and more predictive animal test (i.e., the low volume test).
- Logic analysis of new protocol will optimize test conduct and performance. This is essential for developing new test systems before validation is attempted.
- Critical statistical analysis of test performance new (*in vitro* and *in vivo*) test systems will allow more effective design of validation studies.
- Realization that the data from rabbit eye irritation studies are (at best) rank in nature and serve primarily to categorize compound effects is essential for comparing/evaluating *in vitro* test performance.

DERMAL IRRITATION

This is the second most productive area for the development of *in vitro* tests. All of the points above apply, plus the following:

- Unlike the eye test, it is possible to evaluate multiple test compounds on each animal (generally up to six). With this approach, appropriate rotation and randomization of test sites between animals is essential.

SUMMARY

Statistical methods and approaches to experimental design and interactive analysis of test performance are essential elements in replacing or reducing animal usage in toxicity testing. This use requires more than a testing of hypothesis to identify confidence limits about the difference between groups or results. Rather, the use or carefully considered modification of approaches such as those presented in this chapter is required. This conclusion is based on what should be the central tenet of a biological research philosophy — the ability to continuously critically question what we know and whether or not how we do things is the best way.

REFERENCES

Anderson, S. and Hauck, W. (1983). A new procedure for testing equivalence in comparative bioavailability and other clinical trials, *Commun. Statist. Theor. Meth.*, 12, 2663–2692.

Armitage, P. and Allen, I. (1950). Methods of estimating the LD 50 in quantal response data, *J. Hyg.*, 48, 398–322.

Berkson, J. (1955). Estimate of the integrated normal curve by minimum normit chi-square with particular reference to bioassay, *J. Am. Stat. Assoc.*, 50, 529–549.

Bliss, C. (1935). The calculation of the dosage mortality curve, *Anal. Appl. Biol.*, 22, 134–167.

Bliss, C. (1957). Some principles of bioassay, *Am. Sci.*, 45, 449–466.

Bruce, R. D. and Taulbee, J. D. (1991). Statistical considerations for ocular and dermal toxicology studies, in *Dermal and Ocular Toxicology,* D. W. Hobson, Ed., CRC Press, Boca Raton, FL.

Bruce, R. D. (1985). An up-and-down procedure for acute toxicity testing, *Fundam. Appl. Toxicol.*, 5, 151–157.

Depass, L. R., Myers, R. C., Weaver, E. V., and Weil, C. S. (1984). An assessment of the importance of number of dosage levels, number of animals per dosage levels, sex and method of LD50 and slope calculations in acute toxicity studies, in *Alternate Methods in Toxicology,* Vol. 2 A. M. Goldberg, Ed., Mary Ann Liebert, New York, 139–154.

Finney, D. J. (1971). *Probit Analysis,* 3rd ed., Cambridge University Press, New York.

Finney, D. (1985). The median lethal dose and its estimation, *Arch. Toxicol.*, 56, 215–218.

Gad, S. C. (1989). Statistical analysis of screening studies in toxicology with special emphasis on neurotoxicology, *JACT*, 8, 171–183.

Gad, S. C. and Chengelis, C. P. (1988). *Acute Toxicology Testing Perspectives and Horizons,* The Telford Press, Caldwell, NJ.

Gad, S. C., Smith, A. C., Cramp, A. L., Gavigan, F. A., and Derelanko, M. J. (1984). Innovative designs and practices for acute systemic toxicity studies, *Drug Chem. Toxicol.*, 7, 423–434.

Gad, S. C. and Weil, C. S. (1988). *Statistics and Experimental Designs for Toxicologists,* 2nd ed., The Telford Press, Caldwell, NJ.

Healy, G. F. (1987). Power calculations in toxicology, *ALTA,* 15, 132–139.

Karber, G. (1931). Contribution to the collective treatment of pharmacological serial experiments, *Arch. Exp. Pathol. Pharmacol.*, 162, 480–483.

Krewski, D. and Franklin, C. (1991). *Statistics In Toxicology,* Gordon and Breach Science Publishers, New York.

Morrison, J., Quinton, R., and Reinhert, H. (1968). The purpose and value of LD50 determinations, in *Modern Trends in Toxicology,* Vol. 1, E. Boyland and R. Goulding, Eds., Appleton-Century-Crofts, London, 1–17.

Salsburg, D. S. (1986). *Statistics for Toxicologists,* Marcel Dekker, New York.

Schutz, E. and Fuchs, H. (1982). A new approach to minimizing the number of animals in acute toxicity testing and optimizing the information of the test results, *Arch. Toxicol.*, 51, 197–220.

Thompson, W. and Weil, C. (1952). On the construction of tables for moving average interpolation, *Biometrics,* 8, 51–54.

Waud, D. (1972). Biological assays involving quantal responses, *J. Pharmacol. Exp. Ther.,* 183, 577–607.

Weil, C. (1983). Economical LD50 and slope determination, *Drug Chem. Toxicol.,* 6, 595–603.

Weil, C. (1952). Table for convenient calculation of median effective dose (LD50 or ED50) and instructions in their use, *Biometrics,* 8, 249–263.

Chapter

11

Computer Assisted Drug Dosing

David R. Schneider
Department of Pharmacology
Wayne State University School of Medicine
Detroit, MI

Jill M. Bensie
Drug Dosing Services, Inc.
Southfield, MI

Jim Harlan
Cogitate
Southfield, MI

THE ADVENT OF NEW DRUGS FOR CLINICAL THERAPY

Currently Used Drugs With Therapeutic Actions Often Have Adverse Side Effects

The profession of clinical pharmacology has become, in the past 10 years, a specialty for individuals with expertise in pharmacology, toxicology, and clinical medicine. Individuals expert in clinical pharmacology are called upon to provide a variety of capabilities in multiple practices, including drug dosing services, facts regarding drug interaction profiles and data regarding the therapeutics of an ever increasing number of commonly used drugs. Some of these problems require the use of wide-ranging database services or mastery, e.g., drug interactions, drug toxicity, and particular information about drugs that would be helpful to the elderly, children, or other special cases where

0-8493-8896-1/93/$0.00 + $.50

TABLE 1
Common Drugs Using Therapeutic Drug Monitoring (TDM) Dosing To Achieve Optimal Clinical Effectiveness

Drug classes	Specific drugs
Antiasthma agents	Theophylline
Cardiac drugs	Digoxin, Quinidine, Procainamide, Lidocaine
Antibiotics	Aminoglycosides, Vancomycin
Anticoagulants	Heparin, Warfarin
Antipsychotic agents	Lithium, Mellaril, Haldol
Anticonvulsant agents	Phenytoin, Phenobarbital, Valproic acid, Carbamazepine

the drug(s) is being used. Other services entail knowledge of computer programs, of national database services, and of current information available in the literature. Of particular importance to a successful practice, however, is the knowledge of drug information and customized dosing services related to how a specific drug can be optimally dosed to meet the needs of the individual patient. This problem, termed ''therapeutic drug monitoring'', (TDM), has become increasingly acute and important for two reasons: (1) because there is an ever increasing number of drugs which, even when used according to the guidelines of the Physicians Desk Reference (PDR) or other general information services (Indexes), have such individual characteristics dependent upon the patient that they can not be given optimally by a physician; and (2) because there are an increasing number of efficacious agents coming to market for the treatment of a variety of illnesses with characteristics requiring cautious individual dosing to minimize side effects or toxic conditions or manifestations.

Table 1 summarizes some common clinical agents that, because of a collective pharmacologic property, e.g., high binding to blood proteins, have only a small fraction of their total concentration available for therapeutic action. Because drugs are carried on plasma proteins, (particularly albumin), new or additional drugs taken by the patient can displace drugs that may already be bound. Once a drug is unbound from a serum protein, it is ''free'', and can manifest its actions on other body tissues, e.g., muscles, brain, secretory organs, etc. Thus, if a carefully controlled drug, for example a potent anticonvulsant, is displaced from its ''bound'' serum albumin reservoir, toxic conditions can quickly develop in the patient. To minimize this problem usually requires considering multiple patient variables under the constraints of the physical parameters of the drug in the body. Achieving a consistent end result is not only difficult without a computer program to sort the variables and find solutions through iterative formulas and algorithms, it is nearly impossible. That's why highly protein bound drugs (pharmaceutical agents, chemicals) greatly benefit from TDM programs that individualize the amount of drug administered to a patient.

Today, the use of drug dosing in major medical centers and larger hospital settings has become nearly routine, as physicians, administrators, and insurance companies work together to find the means to reduce health costs while maintaining the standards of care required by their clients. More and more frequently, computer drug dosing programs are required for providing quality health care while saving valuable dollars. Health delivery programs can no longer afford the luxury of unnecessary days of care by skilled personnel while waiting for laboratory results to define the "correctness" of potentially toxic drugs in everyday use. Exacting dosages managed to high tolerances are not only accepted, but required in university and teaching hospitals. And with skillful use of everyday electronics — a FAX machine, a modern 386 or 486 based microcomputer, and the application of artificial intelligence in computer programming — such programs are now available to reach virtually each person that might require such a service, including the physician office, all long-term care and nursing home facilities, and the increasingly expanding home health care programs.

Practical Aspects of Drug Dosing

Computer Dosing Can Lower Medical Costs — Decreased Medication Costs. An obvious use of the TDM technology is to reduce drug costs while providing optimal drug to the patient for the individual clinical therapeutic requirements. In the literature, there are recurring reports documenting and attesting to this concern. Most obvious are the results obtained with the use of aminoglycoside antibiotics (amikacin, tobramycin, and gentamicin) and the related agent vancomycin. These powerful agents are routinely used to combat infections involving highly resistant bacteria such as Streptococci sp., and Proteus sp., organisms that for the most part are nearly untreatable with other agents. It is extremely important to closely monitor the blood concentrations of these particular agents because of their high potential for toxicity to the inner ear and kidney. However, under conditions where TDM is imposed and followed for these agents, it is found that there is always a decrease in drug costs for an individual patient in a specific clinical course. That is because under a refined TDM program designed for these agents, dosing of the drug for maximal effect and minimal toxicity depends only on the individual pharmacological parameters of how the patient handles the drug being administered. Using data from blood tests measuring the serum concentrations at peak and, more important, trough conditions, pharmacologic guidelines are established which precisely relate how the patient being treated handles a known concentration of drug over a specific time period. At this time, known drug interactions can be introduced to the algorithm to further define how the drug will be distributed, metabolized, and excreted from the body. So, unlike general desktop references that point to a minimum effective dose for a wide population range, a refined TDM program can provide specific recommendations to the physician for the individual patient about when and how often the drug should be administered by the clinician.

TABLE 2
Home Health Care Markets In Which Therapeutic Drug Monitoring (TDM) Can Be Effective and Save Money

Long-term care	Nursing homes
	Long-term care facilities such as county agencies
	Hospices
Intermediate care	Home health care
	Short-term hospitals
Short-term care	Home health care
	Walk-in clinics
	Doctor offices

Decreased Care Costs — Less Nursing Care Required. One of the most costly aspects of health care involves the major dollars with which to pay skilled workers. For most illustrations, this category is nearly always designated as the funds required to pay nursing staff. When patients are administered any of the drugs shown in Table 1, there are major concerns regarding drug interactions and, more important, drug toxicity, because of the particular pharmacology involved with each of these agents, even though they reside in entirely different drug classes. Because of the potential for drug interactions and because drug toxicity can develop and become manifest within only a few hours, whenever any of the agents summarized in Table 1 are administered, there is a need for more watchfulness on the part of the nursing staff — time that would otherwise be spent in other nursing care functions or in administrative assignments. Because nursing staff costs represent a fixed expenditure in health care, if a nurse must monitor a patient receiving medications that, if unattended, would be toxic, additional nursing staff must be obtained if a standard of care is to be maintained. Generally, an experienced nurse can monitor and administer care to approximately 10 patients on a given 8 to 12 h shift. If that same nurse must monitor even one critical care patient, thorough charting, nursing notes, and transcribing physician orders will substantially diminish the effectiveness of that nurse to no more than 5 or 6 patients per shift of work

Drug Dosing Can Be Used In Home Health Care Programs. An important new use for TDM is the control of health care costs in areas other than the traditional teaching hospital at a medical center. The fastest growing market in health care is that given to sick people outside of the hospital. These markets are broadly summarized in Table 2.

In most of these markets, virtually all of the agents listed in Table 1 are commonly used, and are administered every day to hundreds of patients. Since most of these care locations tend to have clients who are elderly, caregivers (nursing staff) in such locations will be faced, every day, with multiple problems involving polypharmacy. Because elderly persons take, on average, 6 to 10 prescription medications per day, effectively dosing those drugs that can result in rapid major clinical toxicity presents an important logistical problem.

Because of the burgeoning healthcare costs in hospitals and other agencies, within the past 2 years there has been growth of a true home health care business for tending to the needs of the elderly and other homebound individuals. Generally, this business provides nursing care on a daily or weekly basis, will deliver important drugs such as antibiotics and other agents by i.v. therapy, and will also tend to additional needs including X-rays, laboratory tests, eye care, skin care, and other related needs.

This market is especially capable of defining the needs of persons who may not be critically ill, but are being administered medications defined in Table 1 — e.g., those agents that have a high potential for causing serious toxicity in an uncontrolled state.

Increased Efficacy

Less Underdosing Or Overdosing

Clinical Chemistry Reports and Drugs. In the care of patients from the standpoint of a reference laboratory, an important service lies in providing information to nursing staff and, sometimes, physicians, about how the clinical chemistry or microbiology report can be interpreted. For interpretation about the impact of drugs on specific tests related to clinical chemistry, it is always practical to first review information about the specific test found in standard textbooks of clinical laboratory medicine, e.g., Tietz. More detailed information can be obtained from other sources, including a compilation of information from the literature published by the AACC in 1989, and recently updated (1991). Finally, the latest studies can be searched from a MEDLINE database through most medical libraries.

Microbiology Reports and Drugs. In the area of microbiology, there are only 4 drugs which are of immediate concern: the aminoglycosides (amikacin, gentamicin, and tobramycin), and vancomycin. Each of these agents has the unique characteristic of undergoing renal metabolism and excretion rather than a conventional liver metabolic pathway involving either the cytochrome P-450 system or the synthetic pathways through which glucuronides or sulfatides are formed. Drug dosing via computer algorithms lends itself particularly well to a scenario involving maintaining plasma levels of these agents, and can be extremely effective in adjusting the medication dosage to the patient, to the site of infection and, importantly, to the causative organism of infection. Thus, while it is appropriate to have minimum inhibitory concentrations (MICs) for many antibiotics when agents other than these 4 are considered, use of the aminoglycosides or vancomycin are particularly reserved for certain gram positive or gram negative highly resistant organisms such as Staphylococcus, Proteus, Klebsiella, Enterobacter, Pseudomonas, and

TABLE 3
Error Found In Clinical Laboratory Samples of 14 Dosable Drugs From Nursing Homes, Long-Term Care Centers, Hospitals and Physician Offices

Drug	Total no. of samples	% Error (low + high serum levels)	% Error as subtherapeutic serum level
Anticonvulsants			
Valproic acid	621	53.30	85.80
Phenobarbital	2359	42.10	96.50
Dilantin	6264	62.10	81.80
Tegretol	1190	28.70	85.60
Primadone	326	66.60	100
Cardiac Drugs			
Procainamide	450	64.40	94.5
Digoxin	7661	42.40	91.00
Quinidine	766	68.90	98.50
Antibiotics			
Amikacin	107	86.90	100
Tobramycin	74	40.50	96.70
Gentamicin	381	81.10	97.70
Vancomycin	424	87.50	80.30
Other Agents			
Lithium	3825	39.50	87.60
Theophylline	3997	46.50	92.10

Serratia. For each of the aminoglycosides, and for vancomycin, an effective clinical course requires that the agent be administered at a maximum dose for bactericidal actions, and that there be a sufficient time interval between dosings to allow for drug accumulated in the inner ear endolymph to be reduced to a level that will be non-toxic to the hair cells within the inner ear.

Pharmacokinetic dosing also minimizes potential damage to the renal proximal tubule where excretion of the drug takes place. Vancomycin will accumulate to striking concentrations in a functionally anephric patient, or in conditions of hepatic insufficiency, where it can have a half-life of 6 to 10 d. Importantly, this agent cannot be removed by dialysis.

Examples from Three Different Patient Populations

In conjunction with Preferred Laboratory, Auburn Hills, MI, Dr. Schneider was able to track and follow the clinical course of 14 drugs that fall into the category of agents that must be dosed by computer algorithms for therapeutic levels to be consistently maintained. The results of this study are presented in Table 3. For each of the indexed drugs (Column 1), the percent error in the serum level from a therapeutic window of effectiveness, and including both high and low values as errors is calculated. More important, however, are the tabulated results shown in Column 4: the percentage of error as totaled for all of the subtherapeutic serum levels obtained: e.g., for valproic acid, of 621 samples, 53.3% indicated an error in dosage because the serum level was not within the therapeutic window; of those samples in error, 85.8% had a value lower than the minimum level for therapeutics when using this drug.

As was discussed above, each of the 14 drugs selected for this study share a common characteristic of being highly protein bound. This means, typically, that only 1 to 10% of the total plasma concentration of drug that is measured is "free" or available for pharmacokinetic actions. One extremely important consequence of such high binding is the potential of these agents to be extremely toxic when an overdose situation is encountered. Because of this problem, and because of the numbers of drug interactions with other agents, with disease states, and with the personal physiology of the individual taking the drug, a precise therapeutic concentration of such agents can be more carefully maintained through the use of several computer algorithms working in concert to find a best-fit for the kinetics of the drug in the specified individual. Algorithms are known for most of the individual parameters associated with a specific drug.

It is only when the various algorithms are carefully placed into an integrated design that a computer program can effectively predict and individualize the pharmacokinetics of that agent for the particular clinical problem under consideration.

Table 3 shows the totals for all data, including the average percentage of error in the clinical dosing schedule. Collectively, for all agents considered there was an error of 57.9% ± 0.05% (Mean ± S.E.M.) from the therapeutic range.

Important for this study, however, the overall error was found to be not normally distributed: e.g., with equal or nearly equal error for both toxic and subtherapeutic clinical condition. For each of 14 drugs indexed, the error was highly skewed to the subtherapeutic dosing schedule. Column 4 of Table 3 tabulates the percentage of error found as a subtherapeutic dose. Within all of the agents, profound differences were detected for how each was used in hospitals, in nursing homes or long-term care facilities, and in physician offices. A summary of these differences is briefly discussed below.

Hospitals. In reviewing the use of each of the agents, it was found that hospitals (data from 5 hospitals was pooled for this study) used drugs from each of these categories quite selectively. For example, hospital use of anticonvulsant agents was found to be limited to phenytoin or carbamazepine (tegretol), with only a few instances of primadone use. Valproic acid was rarely, if ever, used in the hospital sample. And, although hospitals used each of the cardiac agents, on a normalized scale (when the use of these agents in nursing homes and doctor offices was considered), hospitals were only minimal users of anticonvulsants. In hospitals, there was a highly dependent use of tobramycin or gentamicin — examples of agents that are probably only minimally effective against *Staph. aureus* and other virulent organisms. More expensive agents known to be more effective against such organisms, e.g., amikacin or vancomycin, were rarely used. Finally, the hospitals used in this survey made use of lithium in a minimal way, if it was used at all, in the treatment of monopolar or bipolar manic illness.

Nursing Home or Long-Term Care Facilities. Use of any of the 14 agents studied in the nursing home or long-term care facility was far more extensive than in the hospital segment. This study included more than 160 nursing home/long-term care facilities. Of the anticonvulsants indexed, only carbamazepine (tegretol) was not extensively used. Nursing homes used each of the cardiac agents, but neither utilized nor understood the importance of computer dosing for quinidine when given by itself or, more probably, with digoxin. Nursing homes relied heavily on tobramycin and gentamicin — like hospitals. However, unlike the hospitals, nursing homes were effective in the use of amikacin as well. There was a very high use of lithium and theophylline in the nursing home/long term care facility division of this study.

Physician Offices. The physician office segment of the study was extremely interesting because of the patterns of drug usage that could be tabulated. For example, although physicians would utilize and dose each of the anticonvulsants, unlike each of the other segments studied, the physicians relied heavily on tegretol as their major agent in this category. Approximately 35% of the use of phenytoin and valproate was concentrated in the physician office segment as well. Because of the i.v. or i.m. use of the antibiotics, physician offices did not contribute to their use in this study.

PHARMACOKINETIC HANDLING OF A DRUG DETERMINES TOXICITY, SIDE EFFECTS, AND THERAPEUTIC EFFICACY

The effectiveness of a drug, once introduced into the body, is highly dependent upon how that drug is controlled and directed by pharmacokinetic forces within the body. Pharmacokinetics, the study of how drugs are interactive within the body, consists of 4 major actions by the body on a drug. These include: (1) how the drug is taken up by the body; (2) how the drug is distributed in the body; (3) how the drug is metabolized by the body; and (4) how the drug is excreted by the body. Each action, while qualitatively independent, is highly dependent on the outcome of the previous action taken by the body on the drug. And since an individual must be considered as a whole when a drug is administered, computer algorithms are required to evaluate and integrate all of the actions by the body when predicting a meaningful outcome for clinical therapeutics.

General Principles of Pharmacokinetics

Absorption

For most therapeutic agents, there is a minimum effective plasma concentration required for clinical efficacy. As a result of this, both the rate and the extent of absorption of a drug can affect its clinical outcome. In many cases, the duration of pharmacologic effect is a function of the length of time the blood concentration is above the minimum effective plasma concentration; the intensity of the effect is usually a function of the height of the blood level curve above the minimum effective concentration.

Distribution

Drug distribution, including the concepts of the volume of distribution (V_d), first pass effect, and half-life ($t_{1/2}$), are integral to the algorithms involved in computer dosing. Calculation of changes in the volume of distribution are most often iteratively made, with a basis for completion being the concentration of drug present in the laboratory sample. Simply described, if the drug is highly lipid soluble, or if the disease state involves conditions where large reservoirs can exist to accumulate drug, the V_d, or the conceptual space that the drug can occupy in the body, will increase and the plasma concentration will decrease accordingly. Calculations of $t_{1/2}$ remain for most agents as the other major factor considered for computer assisted drug dosing. Under most conditions, calculation of the half-life is performed with algorithms which iteratively model a first-order (ln) elimination of the drug. By knowing or extrapolating the peak plasma concentration, and knowing the trough concentration (by laboratory analysis), together with the time interval between these two measures, a first order equation can be found that will fit both ends of the curve. If the half-life increases (e.g., there are substantive changes in the course of the disease process), the trough will be found to occur at a time different than that predicted when dosing was initiated. The algorithm for adjusting half-life will be recalculated and adjusted for a therapeutic range based on the result obtained for the volume of distribution.

Additional parameters such as the first-pass effect (where only a fraction of an oral dose is allowed into the general circulation) are important, but have only limited utility in a drug dosing program. Similarly, how drugs are interactive and how they are individually handled when there are situations of polypharmacy are important only during an initial dosing calculation, or if the various drugs are discontinued. Once the individualized V_d and $t_{1/2}$ have been computed and a rational dosage of medication has been delivered, continued computer dosing will maintain all of the relationships established at an initial dosing protocol.

Metabolism

Drug metabolism has a direct impact on computer dosing algorithms since it is only through drug metabolism that the blood concentration will diminish over time. For most agents, drug metabolism is concurrent with liver metabolism, including both microsomal and non-microsomal pathways. In such cases, drug metabolism is highly correlated with the disease state of the liver, including liver insufficiency in older individuals or those with liver disease. Only rarely are there substantive changes in the metabolic function involving the liver once a protocol of drug dosing is initiated.

For some agents, notably the aminoglycoside antibiotics and vancomycin, metabolism is intimately involved with drug excretion by the kidney. For any agents metabolized and/or excreted by the kidney, it is of primary importance to know the serum creatinine, a measure of kidney function. This analyte, rather than the blood urea nitrogen (BUN), provides an index of clearance of blood through the kidney, and is the major determinant of dosing for these antibiotics.

Excretion

Excretion completes the cycle of drug pharmacokinetics for drugs administered by any route. For many agents, excretion is merely the elimination of a water-soluble metabolic product of the drug that was administered. For others, it is the elimination of the agent without substantial changes in its structure via the kidney. For yet a third class of agents, there is significant metabolism through the liver and into the bowel, where either elimination may occur, or where reabsorption can take place.

Excretion also involves how the body handles products of the primary or initial metabolism of the drug. For many agents, metabolites constitute a substantial fraction of the action of a drug, and metabolic products have significant actions which both mimic and alter the actions of the primary agent.

TECHNOLOGY INVOLVED IN DRUG DOSING

Drug Dosing Request

With the list of drugs being prescribed to patients ever growing, it is impossible for a physician to know all the interactions, toxicities, and complete therapeutics of each drug they might prescribe, even with the use of common reference systems such as the Physicians Desk Reference, Facts & Comparisons, or other drug indices. By looking in such references for drug dosing and specific information, these physicians provide their patients the most prudent care. At the present time, with TDM virtually required for use in large teaching hospitals and universities, younger physicians and those attending Grand Rounds or seminars of CME credit are more familiar with the concepts and the use of TDM data. Older practitioners sometimes have more difficulty in allowing drug dosing to be performed on their patients, or even accepting the results of a drug dosing protocol, because some of the drug

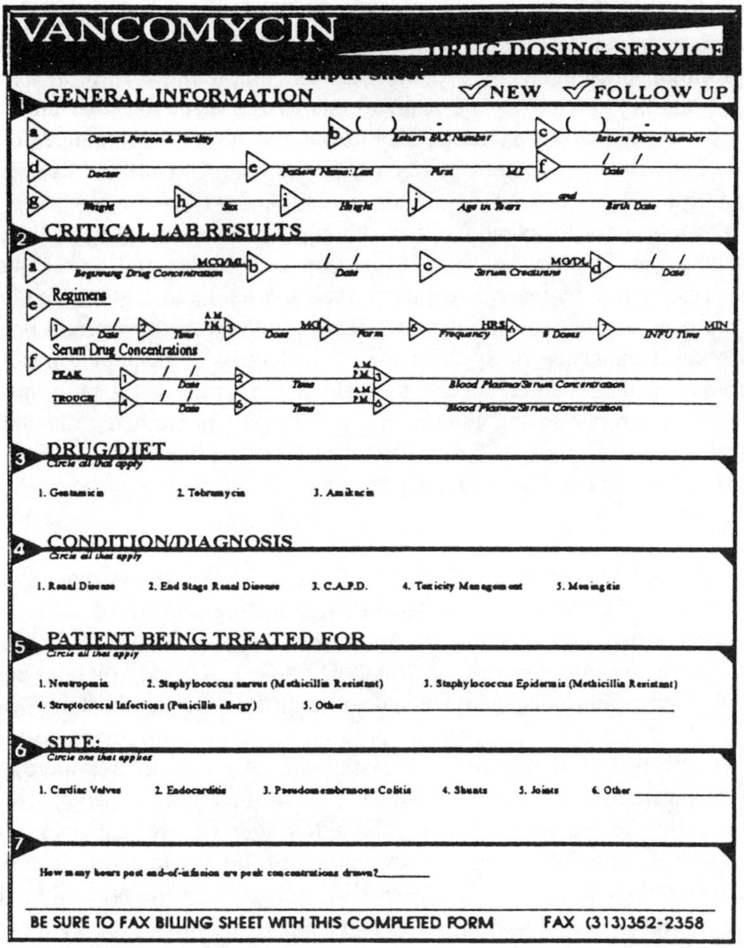

VANCOMYCIN

DRUG DOSING SERVICE

Input Sheet

1 GENERAL INFORMATION ✓NEW ✓FOLLOW UP

a Contact Person & Facility — b ( ) - Return FAX Number — c ( ) - Return Phone Number

d Doctor — e Patient Name: Last First M.I. — f / / Date

g Weight — h Sex — i Height — j Age in Years and Birth Date

2 CRITICAL LAB RESULTS

a Beginning Drug Concentration MCG/ML — b / / Date — c Serum Creatinine MG/DL — d / / Date

e Regimens

1 / / Date — 2 Time A.M. P.M. — 3 Dose MG — 4 Form — 5 Frequency HRS — 6 # Doses — 7 INFU Time MIN

f Serum Drug Concentrations

PEAK 1 / / Date — 2 Time A.M. P.M. — 3 Blood Plasma/Serum Concentration

TROUGH 4 / / Date — 5 Time A.M. P.M. — 6 Blood Plasma/Serum Concentration

3 DRUG/DIET

Circle all that apply

1. Gentamicin 2. Tobramycin 3. Amikacin

4 CONDITION/DIAGNOSIS

Circle all that apply

1. Renal Disease 2. End Stage Renal Disease 3. C.A.P.D. 4. Toxicity Management 5. Meningitis

5 PATIENT BEING TREATED FOR

Circle all that apply

1. Neutropenia 2. Staphylococcus aureus (Methicillin Resistant) 3. Staphylococcus Epidermis (Methicillin Resistant)
4. Streptococcal Infections (Penicillin allergy) 5. Other ______

6 SITE:

Circle one that applies

1. Cardiac Valves 2. Endocarditis 3. Pseudomembranous Colitis 4. Shunts 5. Joints 6. Other ______

7

How many hours post end-of-infusion are peak concentrations drawn? ______

BE SURE TO FAX BILLING SHEET WITH THIS COMPLETED FORM FAX (313)352-2358

Figure 1. Sample drug dosing request input form for vancomycin. The simple form, designed by DDSI, indicates all information needed for an accurate drug dosing regimen to be simulated, with explicit instructions for completing the form found on the back of each. It is particularly important that Section 2 — Critical Lab Results — be correctly filled in. This same format can be repeated for the request forms of all drugs to be dosed, making it "user friendly" for clerical help.

dosing concepts are probably strange or otherwise unfamiliar to how they learned to practice medicine.

The appeal of computer-assisted TDM is its simplicity. All that is required is customary patient information and a FAX machine. Standardized forms are easily produced containing all information needed to be filled in and questions to be answered. A typical drug dosing request input form requesting a consult is shown in Figure 1. The illustrated form is specific for dosing vancomycin and was produced by Drug Dosing Services, Inc., (DDSI) located in Southfield, MI. The form illustrates how patient information, disease interactions, drug interactions, diet requirements, and patient lifestyle can be integrated into a single document. To provide ease in completing the form, detailed instructions are printed on the reverse side of the form for each individual drug. At Drug Dosing Systems, Inc., once the form is completed, clients simply FAX it to the central facility where the computer-driven drug dosing is performed.

It is important to recognize that although the availability of TDM computer programs is growing, they are extremely complex and are not usable at the present time by unqualified personnel. For example, nursing or clerical staff are generally not able to interact successfully with these programs because they require expertise in specific areas, including clinical pharmacology, toxicology, and clinical chemistry, to evaluate the results of the programs and to insure that proper dosing and monitoring of a patient are being maintained. Typical operators are at a Pharm. D, or Ph.D. educational level.

At DDSI, each Drug Dosing Request Form requests the following information:

1. Patient Information — Part 1 of the form (Figure 1) illustrates the input of standard patient data. Patient history and statistical facts are requested, along with data on current drug regimens and the serum drug concentrations of the drug being dosed. It is extremely important to know not only the dose and route of the drug administered, but to also know the corresponding peak and trough serum levels associated with that particular dose of medication. And, in order to predict the pharmacokinetics of the drug, the input must specify when the drug serum levels were obtained. Trough values (the lowest point on the curve) must be drawn just before the next dosing interval; peak levels must be obtained at the proper time intervals for the prescribed drug. In addition to these physical constraints, interacting drugs and conditions, current diagnoses, and the sites of infection, etc. must be included with patient information in order for the most accurate dosing schedule to be generated.
2. FAX Machine — Once patient information has been transcribed, the form is FAXed to the central TDM facility of DDSI. For the first time, doctor and patient need not be in the same city, nor even the same state, as the personnel and operators who are performing the drug dosing. The use of a FAX machine and telephone for additional clarification (if needed) now make TDM available to virtually every physician, hospital, nursing home, and health care facility.

Data Input And Simulation Kinetics Of The Patient

It is very important that the correct patient data is obtained and subsequently entered into the TDM computer program. Drug interactions, diet restrictions, kidney and/or liver dysfunctions, etc. are all correlated from databases built into the program. In the process of simulating the drug kinetics on the patient, such variables will be pulled as needed from look-up tables to optimize the output of the dosing regimen for therapeutic responses. And, should any of the parameters change for a patient over a course of drug dosing, e.g., as a patient becomes less febrile or sick while taking vancomycin, the V_d, the $t_{1/2}$, and the serum creatinine, along with other parameters, can be altered. The new data can be FAXed to the dosing center and entered into the patient's database so that appropriate adjustments will be made in the dosing regimen.

For pharmacokinetic dosing to be effective, there are 3 major indices that must be produced by the program, either from standardized population data, or from the data input for a specific patient. These indices include the volume of distribution of the drug, the half-life of the drug, and the clearance rate.

The Volume of Distribution (V_d) is a measure of the apparent space in the body that is available to contain the drug. These spaces include plasma proteins, e.g., albumin, cellular reservoirs, and fat. V_d defines the relationship between the amount of drug in the body and the concentration (C) of drug in blood or plasma by the equation:

$$V_d = \frac{\text{Total amount of drug in body}}{C}$$

Clearance (Cl) is probably the single most important factor in determining a reasonable drug dosing regimen. It is dependent upon the dose, blood flow, and intrinsic function of the liver or kidneys. In order to obtain steady-state concentrations within the therapeutic range of a drug, the dosing rate must equal the elimination rate. This is defined by the equation:

$$\text{Dosing rate} = \text{Cl} \times C_{ss}$$

where C_{ss} = concentration at steady-state.

Elimination of most drugs follows first-order kinetics, meaning that a constant fraction of the drug is eliminated per unit time. When drug concentrations approach saturation limits, elimination becomes dose-dependent and zero-order kinetics result, where a constant amount rather than a constant fraction of drug is eliminated per unit time. It is essential to know whether the drug follows first- or zero-order elimination kinetics. Although most drugs are first-order, common examples of drugs whose elimination is dose-dependent in the therapeutic range are phenytoin, aspirin, and ethanol.

The half-life ($t_{1/2}$) is the time required for 50% of a drug to be eliminated when first-order kinetics are followed. Half-life is independent of drug concentration and dosage, and is defined by the following relationship between volume of distribution and clearance:

$$t_{1/2} = \frac{0.693\ V_d}{Cl}$$

To provide a therapeutic level of drug to a specific patient, the apparent V_d and $t_{1/2}$ of that agent must be adjusted individually according to a patient's disease state, liver and/or kidney function, concurrent medication, and drug tolerance, together with other factors that might affect an ideal absorption, distribution, and elimination of that drug. Prediction of a dosing regimen is done by analyzing the serum drug concentration points obtained with the current (existing) regimen (or, in the case of an initial dosing, with standardized parameters from population information) using either linear (non-saturated) or non-linear (saturated) curve fitting equations. Non-linear curve fitting is most often performed by employing weighted curve fitting equations, and are referred to as either Simplex or Bayesian weighting formulas. All of these curve fittings make use of standardized formulas found in textbooks of pharmacology. The TDM programs merely allow multiple variables to be processed in parallel, and provide rapid iterative calculations to find the solution(s) with the least error in the predicted curve (from the actual data points entered for the patient).

Output Of Kinetic Simulation

Repeat Of Patient Information

For the drug dosing programs at DDSI, once the dosing regimen has been determined and optimized by the computer protocol, a printout is routinely obtained. In this "hard copy" output, which is FAXed to the physician or pharmacist, are all of the variables relating to the patient information, the current dosing regimen(s) of the patient, the diagnosis of the patient, any concurrent medications the patient is taking, and all of the pertinent serum blood concentrations, etc. originally submitted for the TDM request. Such output is a requirement by DDSI, and serves as a final check, insuring that all input was entered correctly. Included also on the printout is a new "recommended drug dosing regimen", based upon the submitted information.

Graphic Illustration Of Drug Interaction From Simulation

Together with the patient demographics, a graphic illustration of the current and recommended dosing regimens is output for the final record to the physician. A sample of this graph is shown in Figure 2. The graphic includes all serum drug concentration points, together with the simulation of patient response on the recommended dosage. In this manner, the physician

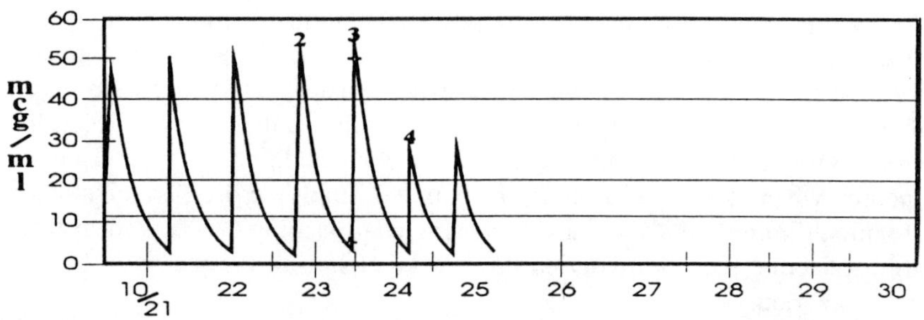

Figure 2. Graphic illustration of drug dosing simulation for vancomycin. The patient was originally dosed at 1200 mg Q 18 h, resulting in a peak serum concentration of 43.5 mcg/ml and a trough of 6.1 mcg/ml. Target therapeutic concentration for vancomycin is 30 mcg/ml, while trough serum concentrations should be below 10 mcg/ml. After simulation, the dosing recommendation was revised to 650 mg Q 12 h, with a simulated peak serum concentration of 29.9 mcg/ml and a trough of 6.2 mcg/ml.

or other health care professional can note that the new regimen is designed to maintain the patient within the therapeutic range of the drug.

Fax To Recipient

More and more, the standardized use of the FAX to convey information between distant locations is understood. For drug dosing applications, the FAX of the output from the drug dosing protocol provides the contact person with nearly immediate information to begin a new dosing regimen. Further, as new serum drug concentrations are obtained or other patient demographics change, upgrading information can be communicated to the dosing central office, entered into the patient's database, and further TDM performed.

MAXIMIZING DRUG ACTIONS

The goal of any drug dosing program is to maximize the therapeutic actions of the drug selected for therapeutics. For most individuals, this goal can only be accomplished by knowing all other conditions of the patient that may affect the action of the drug, including disease states, liver and/or kidney dysfunctions, diet, smoking habits, etc.

However, another important factor in evaluating the response of a drug is the concurrent administration of other drugs. Such considerations are especially prudent when considering drug dosing in an elderly population, since these patients tend to require up to 12 or 15 medications per day to maintain themselves. When a TDM program is used in such a population, mechanisms of drug interactions become extremely important. Moreover, because potential drug interactions are prepared for in the dosing programs, TDM is extremely useful to determine what action(s) must be taken (in dosing other interacting drugs) to prevent or minimize harmful or toxic side effects involving either pharmacokinetic or pharmacodynamic mechanisms. Typically, a clinician will observe pharmacodynamic interactions; however, if such problems are anticipated, TDM will minimize their effect. Generally, drugs with similar pharmacologic effects elicit additive or synergistic responses when given concurrently; responses to multiple drug administration are often diminished if they have opposing pharmacologic effects, or if there are conflicting actions involving drug metabolism, absorption, distribution, or elimination.

Many drug interactions are manifested by their pharmacokinetic effects. The simultaneous use of other drugs may affect the (predominantly) gastrointestinal absorption of the target agent in a number of ways. There is an important distinction between effects on the rate of absorption or effects on extent of absorption. It is not usually clinically important if the absorption rate is decreased. However, subtherapeutic serum drug levels may result if the extent of absorption is decreased.

Competition for protein binding sites in blood (especially on the albumin and alpha-1 glycoprotein fractions) and displacement from tissue binding sites are two mechanisms by which multiple drugs disrupt each other's distribution. The net result of such interactions is often an increase in the "free" concentration of the displaced drug in plasma, which increases the (overall) effect of that drug. Further, the blood or serum concentration of a displaced drug may be increased even when the displacement results from tissue binding sites.

A common problem when there is a practice of polypharmacy is a drug interaction that induces or inhibits the metabolism of all other drugs simultaneously administered. Most often such drug metabolizing interactions are the result of changing the extent or capability of the hepatic microsomal enzyme system (cytochrome P-450 system). Simultaneous use of drugs can also affect renal excretion. Renal elimination is particularly important when dosing aminoglycoside antibiotics or vancomycin because of the changes in drug half-life and the potential for toxic actions on the 8th nerve and the kidney.

Finally, the effects of two drugs which have synergistic toxicity on the same organ are usually greatly enhanced when such agents are used together. Although each drug alone may be at subtoxic serum concentrations, the combined effect will magnify any likelihood of damage. In the evaluation of drug toxicity, it is also well known that even if one drug is not toxic to an organ in and of itself, application of that drug can amplify the toxicity of another agent. The result of this brief overview is to instill the thought that virtually all drugs are susceptible to some type of interaction when taken concurrently with other drugs. And, although not all interactions are clinically significant, many of them make important contributions when determining patient therapy.

APPLICATIONS OF PERSONAL COMPUTERS IN PHARMACOKINETIC DRUG DOSING

Almost all state-of-the-art model personal computers are capable of doing an excellent job in calculating and displaying various drug dosing analyses. Many programs are available that allow the knowledgeable user to run pharmacokinetic analyses that arrive at a recommended drug dosing and infusion regimen. The current state-of-the-art in these software packages is a combination of "expert system" and analytical tool.

An "expert system" is a computer program that incorporates many of the rules that an expert must apply to arrive at a solution or an answer. In the current crop of drug dosing programs, this takes the form of tables of drug and disease interactions. Many drugs and disease states are known to affect the absorption and elimination rates of other drugs, and the computer programs will automatically adjust their formulas to compensate for these factors.

The "analytical tool aspect" of programs for pharmacokinetic drug dosing available at the present time relates to the fact that they require a very knowledgeable operator to interpret and analyze the intermediate results, adjust various parameters, and let the program reanalyze the data to get satisfactory results. In many cases, this expert operator must make several attempts at running the program, each time adjusting variables, to get a result within therapeutic ranges. Nursing or clerical staff are generally not able to interact with these programs successfully. Typical operators are at an educational level of Pharm. D or above.

Given a skilled and knowledgeable operator, all of the available crop of programs are capable of producing valuable recommendations as to drug dosing and infusion schedules. These programs all vary in what drugs they are programmed to support and how new research results are incorporated into the formulas.

Since no one program supports all drugs, a user may be required to learn two or more programs to analyze all drugs of importance to him. This presents a problem in that all these programs are reasonably difficult to navigate through, each gathering data and presenting menus in different styles and sequences. Each also has peculiarities in how data is input and fields are reached or skipped over. Also, changing from one dosing program to another requires exiting one and getting into the next. The historical data differs from one program to another, thus requiring the input of a patient's profile to each dosing program in turn.

These types of complications may explain why drug dosing programs are not being used as much as they should. However, there are several trends in the computer industry that may alleviate these problems and lead to a wider use for pharmacokinetic drug dosing analysis programs.

The first is the wisespread acceptance of Microsoft Windows™ on personal computers. This provides an interface that allows users to enter data and navigate through different programs in a consistent manner. Windows™ also allows the user to switch between many different programs quickly, without exiting and starting new programs in the traditional way. Although no one program can be expected to handle all drugs, under Windows™ it is easier for the operator to learn the different programs and switch between them.

Even more important will be the advent of sophisticated analytical techniques that will allow a clerical level person to produce quality drug dosing results. Although none of the existing crop of programs implement these features now, it can be expected that, as more and more clinicians see the value of pharmacokinetic drug dosing and use the existing programs, in the future there will arise programs that are capable of incorporating much of the knowledge and skills of the Pharm. D, skilled Ph.D. clinical chemist, or clinical Ph.D. pharmacologist into them.

Where the existing programs have already incorporated elements of expert systems, future programs will expand on that and also implement aspects of Artificial Intelligence (AI). An AI based drug dosing program will be expected to work with a clerical level operator to produce the results that currently require the Pharm. D level operator.

AI based systems are implemented through a combination of several techniques. The first is the simple extension of the expert systems technique of incorporating known rules of disease and drug interactions into the formulas for expected absorption, distribution, metabolism, and elimination rates under "ideal" conditions. AI techniques can vary considerably, depending on individual implementations. One general technique that will be implemented and applied to drug dosing solutions is that of "iterative approximations". An iterative approximation technique allows a properly programmed computer to take an educated guess at an infusion time and drug level for a particular patient, to then extrapolate the results, and finally, using hierarchial rules, to adjust the dosage of a particular medication up or down to bring the results into a desired range. By slowly and systematically varying the input parameters and then running the analysis, a result equal to that of the current system used with an expert and experienced operator can be achieved.

Of course, even the most thorough application of AI will encounter situations where unexpected results occur and must be referenced to the clinical Pharm. D or Ph.D. It is expected that such encounters will require that they be handled on an exception basis. The larger bulk of dose analyses for any of the drugs currently used (Table 1), and for those agents yet to be developed, are expected to be able to be processed with an AI program at a clerical level. In such instances, the Pharm. D or Ph.D. will only become involved in the more complex cases.

In summary, although there are many drug dosing programs available today, they require personnel qualified at the Pharm. D level or above.

Sometime in the distant future, the necessity for the expert operator will diminish as the drug dosing programs incorporate ever more capable implementations of expert systems and artificial intelligence. Eventually, the typical operator will be clerical, and the expert operator only necessary on complex cases. These new generation computer programs will be run with the consistent user interface found in Windows™, and similar systems will guarantee that the use of computer pharmacokinetic drug dosing programs will become the standard practice instead of the exception that they are today.

Chapter 12

Perspectives on the Impact of Biotechnology to Increase Drug Efficacy and Safety*

Harry Salem
Life Sciences Department
Chemical Research Development and Engineering Center
Aberdeen Proving Ground, MD

Steven I. Baskin
Pharmacology Division
U.S. Army Medical Research Institute of Chemical Defense
Aberdeen Proving Ground, MD

Man has always had the driving desire and need to medicate himself. From simple plant and animal parts applied topically in crude form locally on the skin, taken orally by chewing, swallowing, or drinking as a tea, or by smoking. With the advent of pharmacy, advances have progressed in the preparation of medicines such as extractions, galenicals, elixirs, syrups, injectables and drops as well as aerosols.

We have come a long way in therapy from crude botanicals through galenicals, alkaloids, antibiotics, and synthetic drugs. In addition to the ad-

*The views of the authors do not purport to reflect the position of the Department of Defense. The use of trade names in this report does not constitute an official endorsement or approval of the use of such commercial hardware or software. This report may not be cited for purposes of advertisement.

Requests for reprints should be sent to:
Commander
U.S. Army Medical Research Institute of Chemical Defense
ATTN: SGRD-UV-PB/Dr. S.I. Baskin
Aberdeen Proving Ground, MD 21010-5425

0-8493-8896-1/93/$0.00 + $.50

vances made isolating and preparing specific active ingredients, the methods of administration have become more sophisticated from chewing bark, leaves and berries, through liquids, tablets, capsules, injectables, suppositories, aerosols, wafers, ocular inserts, ointments, salves, and dermal patches. Long-acting and timed release medications are also available. These developments not only serve to make medications pharmaceutically elegant, but to make them more palatable and easy to take or administer. In addition, these preparations were developed to maximize bioavailability and thus efficacy and safety.

This symposium and publication attempt to describe some of the advances in biotechnology which will contribute significantly to the objectives of increasing efficacy and safety of therapeutic agents, as well as targeting the specific protein receptors.

In addition to what was discussed at the symposium, recent literature suggests that a number of biotechnological advances are coming to fruition that will reduce drug toxicity (Weiss, 1988). The editors do not claim to describe all of these recent breakthroughs but attempt to acquaint the readers with some and provide a listing of a portion of the sources that could provide information on these potentially useful products.

Non-invasive techniques have been recently developed to directly monitor blood glucose. A light beam is passed through the finger, the light is processed and can provide blood glucose measurements (Amato, 1992). Development of non-invasive methods such as these have the potential to reduce drug toxicities. These methods are extensions of similar approaches which were described 33 years ago (Bilby and Salem, 1959).

It has been claimed that a small ''targetor'' moiety together with cholesterol will allow large molecules to pass the blood brain barrier and thus allow a protein to be effective in diseases which act at different brain regions. Addition of 1,4-dihydrotrigonellinate may allow peptides to enter the brain. These compounds may be able to treat a wide myriad of central nervous system diseases such as Parkinson's disease and certain types of severe pain (Borman, 1992). Lee (1992) has claimed that he can provide greater oral delivery of peptides by increasing the lipophilicity of drugs.

Greater understanding of how genetic information is transferred and introduced into cells and how current genes can be modified has given rise to practical approaches to human gene therapy. New techniques could make it possible to modify genetic information in order to treat disease that will modify fundamental cellular coding information (Miller, 1992). Examples of defective genes and their diseases can include: adenosine deaminase (immunodeficiency) LDL receptor (hypercholesterolemia); Factor IX and VIII (hemoglobinemia); glucocerebrosidase (Gaucher's disease); alpha 1-antitrypsin (emphysema); phenylalanine hydroxylase (phenylketonuria); beta-globin (thalassemia) and CD-18 (leukocyte adhesion deficiency).

Progress in other diseases such as in cystic fibrosis should allow us to find a treatment for this genetically related disease. A single amino acid deletion in the cystic fibrosis transmembrane conductance regulator (CTTR) is the mutation most frequently responsible for cystic fibrosis (Armstrong, 1992). Repair of CTTR by use of human genetic engineering may allow for the repair of this as well as other genetic diseases.

Transdermal patches are being used as new forms of drug delivery that could offer improvement over older dosage forms. Patches containing medication are taped to the skin. The medication is released and absorbed through the skin into the systemic circulation to reach the target organs. Transdermal patches were originally introduced to deliver anti-motion drugs (see Chapter 6) and are now used as delivery systems for treating heart disease and as an aid to stop smoking by relieving the nicotine craving. Other drugs under consideration for this method of administration include analgesics and hormone combinations as contraceptives to prevent pregnancies. A second generation enhancement of the transdermal patch is the microsponge. These are synthetic, highly porous sponges consisting of a complex network of miniature tunnels and pores which can be filled with therapeutic agents (or cosmetics). The microsponges can be sprinkled on as a powder, and each time the skin is rubbed with the powder, the drug (or cosmetic) is released (squeezed out of the sponge) and absorbed into the skin (Weiss, 1988). It has been estimated that one gram of microsponge contains approximately 240,000 miles of pore length which can be altered in size. The sponges can be designed to release their contents at predetermined temperatures, moisture contents, and other conditions. Another variation of transdermal drug administration is the use of drug impregnated contact lenses. Antibiotics and drugs for treating glaucoma are under consideration for this type of delivery system. Another approach in the treatment of glaucoma is the use of a small drug saturated, biodegradable polymer placed under the eyelid (Ocusert, see Chapter 6).

Traditional drug administration for the therapy of glaucoma is the use of eye drops. This is a cumbersome and inefficient drug delivery system. Approximately 5 to 10% of the therapeutic agent is absorbed into the eye where it is required; the rest is absorbed systemically where it can cause serious adverse effects. Approximately 20 years ago, it was found that ophthalmic aerosols (Salem and Ellison, 1973) provided improvement when compared to drops for the treatment of glaucoma. It was demonstrated that approximately 10% of the dose delivered by liquid eye drops is required to produce the same drop in intraocular pressure when delivered as an aerosol. The same technique is applicable for the treatment of other ocular conditions such as allergies and infections.

Another biodegradable polymer implant is being developed to treat brain tumors. It is a 1-in. diameter implant (small wafer) with layers of slowly dissolving plastic containing anti-cancer drugs. The plastic biodegrades on contact with brain fluid, slowly releasing the contained anti-cancer drugs killing malignant cells (see Chapter 4). This technique may also be applicable in patients with Parkinson disease using dopamine as the slowly released drug.

Studies are in progress in which insulin combined with a sugar sensitive enzyme inside small polymer wafers are implanted into diabetic rats. When the blood sugar level rises in these rats incapable of producing insulin, the sugar sensitive enzyme is activated causing a slight increase in acidity which increases the solubility of the combined insulin increasing its availability. When the blood sugar level drops, the acidity decreases, slowing the release of insulin. Thus, insulin is released only when it is needed in response to increased blood sugar levels. This biofeedback control offers great hope for the future in this and other diseases (Weiss, 1988).

Insulin and other drugs requiring rapid bioavailability or that are susceptible to degradation by digestive enzymes are being investigated for delivery by nasal spray (Lee and Sciarra, 1976; Waldholz, 1984). Permeation enhancers, which are mild detergents, make therapeutic proteins more soluble and enhance systemic drug absorption through the nasal mucosa. Large protein molecules like insulin cannot be absorbed through the nasal mucosa by themselves. In addition to insulin, permeation enhancers are being considered for nasal delivery of growth hormone, drugs for blood pressure control, contraceptives, and for reproductive hormone dependent disorders.

Recently, there has been controversy related to the use of contraceptive implants. Norplant capsules implanted under the skin of a woman's arm can prevent pregnancy for up to five years. Another implant, Capronose, which never has to be removed, can prevent pregnancy for at least 18 months. Vaginal rings release low doses of contraceptive drugs for 3 months and can be removed at any time. Other procedures for contraceptive drug delivery include injectable microspheres and microcapsules. When injected, contraceptive drugs are slowly released into the blood stream and can prevent pregnancy for up to six months. Microspheres are also being used as nanocapsules to deliver insulin by oral administration. Nanocapsules are microscopic spheres of biodegradable polymer originally developed as tissue glue for use in surgery. The spheres, when taken orally in colloidal solution, protect insulin from gastric digestion permitting intestinal absorption where they release the insulin systemically to lower blood sugar levels (Weiss, 1988).

The use of lipid envelopes for carrying drugs for stabilization to allow for passage as a form of pharmaceutical delivery system (liposomes) are currently being developed by many laboratories (Talsma and Crommelia, 1992). Certain compounds, when prepared in a liposome preparation, showed higher efficacy and lower risk of intoxication than control formulations. These formulations could be particularly useful under conditions when products may be unstable in a biological matrix (i.e., proteins) (Talsma and Crommelia, 1992).

Liposomes, either synthetic or of natural ingredients such as egg yolks and soy beans, are tiny fat-soluble spheres with aqueous interiors. Water soluble drugs are bound inside and fat soluble ones in the fatty membrane. The drugs become available when the liposome dissolves or is ingested by a cell. Applications of this type of drug administration include anti-cancer therapy (doxorubicin). Doxorubicin is a highly toxic drug which can cause irreversible heart muscle damage. However, when administered in liposomes, its cardiac toxicity can be reduced by 80%. Another potential anti-cancer drug, cytochalasin B, when administered in liposomes has a reduced toxicity.

Liposome time-release methodology might also be applicable for anesthetic agents. Methoxyflurane, a long acting anesthetic has been prepared in liposomes or lecithin and sugar water. When injected, the water soluble methoxyflurane is gradually released over several days.

An exciting application of liposomes is to target specific receptors. Attaching liposomes to monoclonal antibodies guarantees that the liposomes will bind only to specific target cells where the liposomes are engulfed, digested, and their anti-cancer contents are released. Liposomes can also be used as vaccines such as the one designed to control a disease fatal to chickens.

Another challenge in drug delivery is to specifically target DNA inside cells to permanently change the cell's genetic predisposition to accommodate the individual's medical needs. This type of gene expression regulation could, for example, cause liver cells to excrete insulin when pancreatic cells fail to do so. This type of research has already been successful in experiments where mice were injected with pieces of DNA encapsulated in antibody coated liposomes engineered to attach to mouse cells. These cells ingested the liposomes where they released their load of DNA which was then incorporated into the host cell DNA. There it stimulated the production of an enzyme that the cells had previously been unable to make.

Dr. Steven Rosenberg (1990) has pioneered a new treatment for cancer called adaptive immunotherapy or cell transfer therapy. The goal is to selectively affect cancer cells without affecting normal cells, thus decreasing toxicity. This treatment involves removing immune system cells from a patient, modifying them to make them more active against the patient's specific illness by separating the cells by enzyme treatment and cultured with interleukin-2 (tumor infiltrating lymphocytes). This causes the lymphocytes to increase rapidly and attack cancerous cells in the culture. Eventually, the lymphocytes completely replace the cancer cells. The tumor infiltrating lymphocytes, following at least a months incubation, are infused into the patient along with interleukin-2 (Jones, 1990). This treatment has resulted in some cancer remissions in patients.

Studies are being considered at the University of Michigan where direct gene injections will be made into tumors with the objective of destroying them. HLA-B7 genes mixed with fat soluble particles (liposomes) will be injected directly into skin tumors. The immune system will recognize cells containing HLA-B7 as foreign and will mount a defense against them. This should stimulate the immune system to recognize other unmodified tumor cells and attack them.

New techniques are also important in developing diagnostic procedures. Prostate cancer (diagnosed in 132,000 American males per year with 34,000 fatalities) is now being detected earlier by using a blood test that measures prostate specific antigen (PSA). PSA levels (often elevated in men with prostate cancer) can distinguish cancer from benign prostatic hyperplasia.

A new test may also help in the early detection of colon cancer. It can detect a mutant gene linked to both colon cancer and premalignant polyps by analyzing stool samples. This approach hopefully will be expanded to other genes linked to disease as well as to other forms of cancer. Colon cancer is the world's third most common type of cancer accounting for 20% of U.S. cancer deaths (58,000). Approximately 156,000 cases develop yearly. Stool samples are purified and examined for defective genes (Marx, 1992). This procedure will probably be expanded to use sputum as a potential diagnostic procedure for lung cancer and urine for bladder cancer.

DNA typing which compares the genetic pattern of a biological sample may be used to identify or exonerate a suspect. A report by the National Academy of Sciences provides strong support for the use of DNA fingerprinting in legal proceedings and argues that rigorous controls should be incorporated into the procedure (Koshland, 1992; Neufeld and Colman, 1990).

A new class of antibiotics monolactams, has been developed by the Squibb Institute for Medical Research (Kripalani et al., 1984). One of these monolactams, Aztreonam, taken orally for diarrhea with only 1% absorbed lessens the chance of side effects particularly for pregnant women and children and patients with renal and hepatic dysfunction.

There is substantial growth of products and sales by the emerging biotech pharmaceutical companies (Thayer, 1992). These new products are only some of the exciting breakthroughs that new technologies offer in the future in terms of more effective therapy and reduction in drug toxicities.

REFERENCES

Amato, I. (1992). Race quickens for non-stick blood monitoring technology, *Science,* 258, 892–893.

Armstrong, J. (1992). Another protein out in the cold, *Nature,* 358, 709–710.

Bilbey, D. L. J. and Salem, H. (1959). Assessment of the phagocytic activity of the macrophage system, *Nature,* 184, 992–993.

Blair, K. (1992). Life isn't cheap, *Michigan Today,* Oct. 4–6.

Borman, S. (1992). Blood-brain barrier bridged by molecular packaging technique, *C&EN,* 70, 57–58.

Borman, S. (1992). Role of computers in drug discovery analyzed, *C&EN,* 70, 59.

Bowie, L. (1992). Seeking nature's cures, *Baltimore Sun,* 27 Sept. 1B.

Collins, F. S. and Wilson, J. M. (1992). A welcome animal model, *Nature,* 358, 708–709.

Dibner, M. D. (1992). U.S. biotechnology and pharmaceuticals, *BioPharm,* 5, 24–28.

Fisher, L. M. (1992). New drugs by process of elimination, *New York Times,* 6 Oct. 1D.

Glanz, J. (1992). Pulse quickens in search for blood substitute, *R&D Magazine,* 34, 54–57.

Glanz, J. (1992). Glycobiologists prove the biochemistry of life is sweet, *R&D Magazine,* 34, 62–64.

Gershon, D. (1992). Frustrated by regulatory delays at FDA, biotech companies warm to user fees, *Nature,* 358, 616.

Jones, R. R. (1990). Scientist of the year, *R&D Magazine,* 32, 58–62.

Koshland Jr., D. E. (1992). DNA fingerprinting and eyewitness testimony, *Science,* 256, 593.

Kripalani, K. J., Singhvi, S. M., Weinstein, S. H., Everett, D. W., Bathala, M. S., Dean, A. V., Ita, C. E., Lawrence, L., and Meeker, F. S., Jr., (1984). Disposition of [14C] aztreonam in rats, dogs and monkeys, *Antimicrob. Agents Chemother.,* 26, 119–126.

Lee, V. H. L. (1992). Oral route of peptide and protein drug delivery, *Bio Pharm.,* 5, 39–50.

Lee, S. W. and Sciarra, J. J. (1976). Development of an aerosol dosage form containing insulin, *J. Pharm. Sci.,* 65, 567–572.

Lowe, K. C. and Washington, C. (1992). Alternative blood substitutes, *Nature,* 358, 717–718.

Marx, J. (1992). Test could yield improved colon cancer detection, *Science,* 256, 32.

Miller, A. D. (1992). Human gene therapy comes of age, *Nature,* 357, 455–460.

Neufeld, P. J. and Colman, N. (1990). When science takes the witness stand, 262, 46–53.

Rosenberg, S. A. (1990). Adoptive immunotherapy for cancer, *Scien. Amer.,* 262, 62–69.

Robbins-Roth, C. (1992). Matching your science to the market, *Biopharm,* 5, 18–22.

Salem, H. and Ellison, T. (1973). Pharmacologic action and tissue distribution of epinephrine bitartrate in the rabbit eye, *Ann. Ophthalmol.* 5, 417–426.

Talsma, H. and Crommelin, D. J. A. (1992). Liposomes as drug delivery systems, Part 1: preparation, *BioPharm,* 5(8), 36–47.

Talsma, H. and Crommelin, D. J. A. (1992). Liposomes as drug delivery systems, Part 2: characterization, *BioPharm,* 5(9), 38–41.

Thayer, A. (1992). Biotech industry revenues top $8 billion, *C&EN,* 70,48.

Weiss, R. (1988). Welcome to the space age pharmacy, *Washington Post,* 3 July.

Waldholz, M. (1984). Nose spray may offer means of administering new drugs, *Wall Street Journal,* Jan. 6.

INDEX

D

M

N

O

P

U

V

W